Atsushi Mizuike

W0080460

Enrichment Techniques for Inorganic Trace Analysis

With 41 Figures and 49 Tables

Springer-Verlag
Berlin Heidelberg NewYork 1983

Chemical Laboratory Practice

Editors

F. L. Boschke, W. Fresenius, J. F. K. Huber, E. Pungor,
G. A. Rechnitz, W. Simon and Th. S. West

Prof. Dr. Atsushi Mizuike, Faculty of Engineering, Nagoya University, Chikusa-ku, Nagoya 464, Japan

Editors

Dr. Friedrich L. Boschke, Springer-Verlag, Postfach 105 280, D-6900 Heidelberg 1, FRG

Prof. Dr. Wilhelm Fresenius, Institut Fresenius, Chemische und Biologische Laboratorien GmbH, Im Maisel, D-6204 Taunusstein 4, FRG

Prof. Dr. J.F.K. Huber, Institut für Analytische Chemie der Universität Wien, Währinger Straße 38, A-1090 Wien, Austria

Prof. Dr. Ernö Pungor, Institute for General and Analytical Chemistry, Gellért-tér 4, H-1502 Budapest XI, Hungary

Prof. Garry A. Rechnitz, Dept. of Chemistry, Univ. of Delaware, Newark, DE 19711, USA

Prof. Dr. Wilhelm Simon, Eidgenössische Technische Hochschule, Laboratorium für Organische Chemie, Universitätstraße 16, CH-8092 Zürich, Switzerland

Prof. Thomas S. West, Macaulay Institute for Soil Research, Craigiebuckler, Aberdeen AB9 2QJ, U.K.

ISBN-13: 978-3-642-68856-0   e-ISBN-13: 978-3-642-68854-6
DOI: 10.1007/978-3-642-68854-6

Library of Congress Cataloging in Publication Data.
Mizuike, Atsushi, 1926 –
Enrichment techniques for inorganic trace analysis. (Chemical laboratory practice; 19 –
Anleitungen für die chemische Laboratoriumspraxis; Bd. 19)
Bibliography: p.; includes index.
1. Trace elements--Analysis. I. Title. II. Series: Chemical laboratory practice; 19.
QD139.T7M49 1983; 545; 82-19430

This work is subject to copyright. All rights are reserved, whether the whole or part of the material is concerned, specifically those of translation, reprinting, re-use of illustrations, broadcasting, reproduction by photocopying machine or similar means, and storage in data banks. Under § 54 of the German Copyright Law where copies are made for other than private use, a fee is payable to Verwertungsgesellschaft Wort, Munich.

© Springer-Verlag Berlin, Heidelberg 1983
Softcover reprint of the hardcover 1st edition 1983

The use of registered names, trademarks, etc. in this publication does not imply, even in the absence of a specific statement, that such names are exempt from the relevant protective laws and regulations and therefore free for general use.

Preface

The significant role of trace elements present at the $\mu g/g$ (10^{-6} g/g), ng/g (10^{-9} g/g) and pg/g (10^{-12} g/g) levels in geological, biological, environmental and industrial materials has increasingly been recognized in science and technology. To detect and determine trace elements, we usually use modern optical, electrochemical and nuclear analytical techniques. Although most of them are highly sensitive and selective, preliminary enrichment techniques are required to extend the detection limits, improve precision and accuracy of analytical results, and to widen the scope of the determination techniques. About two decades ago, I wrote a chapter "Separations and Preconcentrations" in "Trace Analysis: Physical Methods" edited by Prof. G.H. Morrison (Wiley-Interscience, New York, 1965). Since then, the progress in this field has been remarkable.

This monograph is intended as *a laboratory book directly applicable to the practice,* but is not a so-called "cookbook" which offers detailed laboratory instructions. I hope this book is useful for all analysts solving problems in inorganic trace analysis and appreciating the applicability and limitations of enrichment techniques combined with instrumental determination techniques.

In three introductory chapters, general aspects and control of contamination and loss are discussed. The following eight chapters deal with enrichment techniques based on volatilization, liquid-liquid extraction, selective dissolution, precipitation, electrochemical deposition and dissolution, sorption, ion exchange, liquid chromatography, flotation, freezing and zone melting. The final two chapters are devoted to special enrichment techniques used in trace analyses of natural waters and gaseous samples.

Finally, I wish to express my appreciation to my collaborators for their help in preparing this book, especially to Dr. Masataka Hiraide for literature survey, proofreading and valuable discussion, to Ms. Atsumi Kato for typing the manuscript, and to Mr. Tomokazu Tanaka for preparing the line drawings. Without their assistance, this book would never have appeared.

August 31, 1982　　　　　　　　　　　　　　　　　　　　　　　Atsushi Mizuike
Nagoya, Japan

Table of Contents

1	**Introduction**	1
1.1	Inorganic Trace Analysis in Science and Technology	1
1.2	The Role of Enrichment Techniques in Inorganic Trace Analysis . . .	2
2	**General Aspects of Enrichment Techniques**	5
2.1	Trace Recovery	5
2.2	Enrichment Factor	6
2.3	Contamination	6
2.4	Simplicity and Rapidity	7
2.5	Sample Size	8
3	**Control of Contamination and Loss**	9
3.1	Airborne Contamination	9
3.1.1	Clean Rooms	10
3.1.2	Clean Hoods or Benches	12
3.1.3	Other Means to Reduce Airborne Contamination	12
3.2	Contamination and Loss Due to Apparatus	13
3.2.1	Selection of Materials	13
3.2.2	Surface Treatment	16
3.2.3	Cleaning of Containers	16
3.3	Contamination Due to Reagents	17
3.3.1	Selection of Commercial Reagents	17
3.3.2	Preparation of High-Purity Reagents in Analytical Laboratories	17
3.4	Other Sources of Contamination and Loss	20
4	**Volatilization**	21
4.1	Volatilization from Solutions	22
4.1.1	Volatilization of Trace Elements from Solutions	22
4.1.2	Volatilization of the Matrix from Solutions	22
4.2	Volatilization from Solid and Molten States	27
4.2.1	Volatilization of Trace Elements from Solid and Molten States	27
4.2.2	Volatilization of the Matrix from Solid and Molten States	28
5	**Liquid-Liquid Extraction**	30
5.1	General Procedures	30
5.1.1	Batch Extraction	30
5.1.2	Continuous Extraction	33
5.1.3	Countercurrent and Chromatographic Extractions	33

5.1.4	Backwashing	34
5.1.5	Stripping	35
5.2	Extraction of Metal Chelates	35
5.2.1	Chelate Extraction Systems	35
5.2.2	Equilibria in Chelate Extraction Systems	37
5.2.3	Masking	41
5.2.4	Synergism	42
5.2.5	Coextraction	42
5.2.6	Extraction Rate	43
5.2.7	Chelate Extraction of Trace Elements	43
5.2.8	Chelate Extraction of Matrix Elements	44
5.3	Extraction of Ion Pairs	46
5.3.1	Ion-Association Systems	46
5.3.2	Ion-Association Extraction of Trace Elements	50
5.3.3	Ion-Association Extraction of Matrix Elements	50
5.4	Special Extractions	50
5.4.1	Three-Phase Extraction	50
5.4.2	Homogeneous Extraction	50
5.4.3	Extraction with Molten Organic Compounds	51
5.4.4	Extraction of Trace Elements from Nonaqueous Samples	51
6	**Selective Dissolution**	52
6.1	Selective Dissolution of the Matrix	52
6.2	Selective Dissolution of Trace Elements	53
7	**Precipitation**	56
7.1	Precipitation of Matrix Elements	56
7.1.1	Coprecipitation Phenomena	57
7.1.2	Application of Matrix Precipitation	60
7.2	Precipitation of Trace Elements	61
7.2.1	Carrier Precipitation	61
7.2.2	Application of Carrier Precipitation	62
8	**Electrochemical Deposition and Dissolution**	67
8.1	Electrodeposition on Solid Electrodes	67
8.2	Electrodeposition on Mercury Cathodes	69
8.2.1	Deposition of Trace Elements	70
8.2.2	Deposition of Matrix Elements	73
8.3	Spontaneous Electrochemical Deposition	73
8.4	Anodic Dissolution	74
9	**Sorption, Ion Exchange and Liquid Chromatography**	75
9.1	General Procedures	75
9.1.1	Batch Operation	75
9.1.2	Filtration through a Permeable Sorbent Disk	75
9.1.3	Column Operation and Chromatography	76
9.2	Separation with Ion Exchange Resins	79
9.2.1	Ion Exchange Resins	81
9.2.2	Ion Exchange Reactions and Equilibria	82

9.2.3   Sorption of Trace Elements on Ion Exchange Resins   83
9.2.4   Removal of Matrix Elements by Sorption on Ion Exchange Resins . .   88
9.2.5   Sorption of Matrix and Trace Elements on Ion Exchange Resins
        Followed by Chromatographic Elution   88
9.3   Separation with Cellulosic Exchangers   88
9.4   Separation with Polyurethane Foams   90
9.5   Separation with Miscellaneous Organic Sorbents   90
9.6   Separation with Activated Carbon   92
9.7   Inorganic Ion Exchangers   93

10   Flotation   94
10.1   General Procedures   94
10.2   Carrier Precipitation Followed by Flotation   95
10.2.1 Important Experimental Factors   95
10.2.2 Applications   97
10.3   Ion Flotation   98
10.3.1 Important Experimental Factors   98
10.3.2 Applications   99

11   Freezing and Zone Melting   100
11.1   Freeze Concentration of Dilute Aqueous Solutions   100
11.2   Enrichment of Impurities in Solids by Zone Melting   101

12   Enrichment Techniques in Water Analysis   103
12.1   Separation Based on the Particle Size and Density   104
12.1.1 Filtration and Ultrafiltration   104
12.1.2 Dialysis   104
12.1.3 Gel Filtration   104
12.1.4 Centrifugation   105
12.2   Separation Based on Chemical Reactivity   105
12.2.1 Volatilization   105
12.2.2 Liquid-Liquid Extraction   106
12.2.3 Carrier Precipitation   106
12.2.4 Electrodeposition   107
12.2.5 Sorption, Ion Exchange and Liquid Chromatography   107

13   Enrichment Techniques in Gas Analysis   108
13.1   Separation of Particles   108
13.2   Separation of Gaseous Trace Constituents   109

        Literature   110

        Appendix   126
A.1   Solvents   126
A.2   Masking Agents   127
A.3   Ion Exchange Data   129

        Index of Abbreviations and Symbols   139

        Subject Index   142

1 Introduction

Inorganic trace analysis or trace element analysis is defined as the *determination of trace elements at concentrations below about 100 μg/g* in inorganic and organic samples. At present, trace elements even at the ng/g (10^{-9} g/g) and pg/g (10^{-12} g/g) levels can be determined with satisfactory accuracy and precision by using proper analytical techniques that, however, involve a number of difficult problems. The difficulties arise mainly from extremely low concentrations of trace elements in various matrices, not from absolute quantities of the trace elements.

1.1 Inorganic Trace Analysis in Science and Technology

Trace elements in terrestrial materials such as the atmosphere, ground, river, lake and sea waters, soils, minerals and rocks, and in cosmic materials such as meteorites, soils and rocks on the luner surface have extensively been analyzed to obtain invaluable information in geochemical and cosmochemical studies. The role of trace elements in biological systems is very complicated. There are a number of essential, beneficial, harmful or toxic trace elements for plants and animals. Most of the essential trace elements whose deficiency gives rise to various diseases are toxic to plants and animals if present in excessive amounts, and the optimum concentration ranges are relatively narrow for some elements. Therefore, the atmosphere, drinking water, soils, plants, animal and human diets, and animal and human blood, urine and tissues are frequently analyzed for trace elements in biological, agricultural and medical sciences as well as in connection with environmental problems. Inorganic trace analysis is also very important in physical sciences and industry. The presence of trace impurities in materials such as high-purity metals, semiconductors and glasses has an important influence on electrical, magnetic, mechanical, nuclear and optical properties as well as on chemical resistivity. Impurities in raw materials such as petroleum and ores may cause troubles during manufacturing processes such as poisoning of catalysts and deterioration of production efficiency. Some trace elements in industrial effluent gases and waters are sources of environmental pollution. Other fields of application of inorganic trace analysis include criminology and archaeology.

To investigate the synergetic action and correlation of trace elements in high-purity materials, biological and environmental samples, etc., analytical results are necessary for the extreme variety of trace elements that may be present in the sample. If available, simultaneous multielement determination techniques are suitable for this purpose, because time, labor, samples and reagents required for the analysis are minimized. Information on chemical forms as well as distribution of a trace element in the sample is also frequently required. As an example, the determination of the chemical forms of trace heavy metals in natural waters gives useful information in studies of geochemistry, environmental problems, biological effects of trace elements, and water treatment.

1.2 The Role of Enrichment Techniques in Inorganic Trace Analysis

The general scheme of inorganic trace analysis is shown in Fig. 1. To ensure sufficient precision and accuracy of analytical results in inorganic trace analysis, great care must be taken to minimize loss of the desired trace elements (trace elements to be determined) and contamination from external sources during the whole analytical process from the sample collection through the determination [1]. In addition, interferences from unexpected trace inorganic and organic substances coexisting in the sample may cause biases in analytical results. These problems generally become critical for trace elements at the concentration levels below 1 $\mu g/g$.

Because of the low concentrations and the small absolute amounts of the desired trace elements, *highly sensitive and selective determination techniques* listed in Table 1 are generally employed. The absolute and relative (concentration) detection limits (expressed in absolute amount and in concentration) as well as the selectivity of each technique vary greatly with elements, matrices, analytical instruments, reagents, and various experimental conditions. Most of the techniques listed in Table 1 have absolute detection limits in nanogram (10^{-9} g) or picogram (10^{-12} g) ranges for many elements with adequate selectivity. In favorable cases, the absolute detection limits for some elements are in the femtogram (10^{-15} g) range or lower.

Direct applications of these determination techniques are, however, frequently impossible, difficult or undesirable,

(1) when the concentrations of the desired trace elements are below the relative detection limits of the determination technique,

(2) when substances which interfere with the determination exist in the sample,

(3) when the sample is highly toxic, radioactive, or expensive to be wasted,

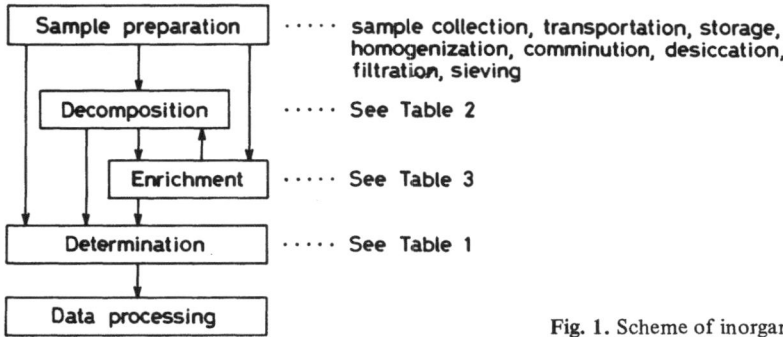

Fig. 1. Scheme of inorganic trace analysis

Table 1. Determination techniques used in inorganic trace analysis

---

Molecular absorption spectrometry — Ultraviolet. Visible. Infrared.
Luminescence spectrometry — Ultraviolet (Fluorometry). X-ray.
Photoacoustic spectrometry
Atomic absorption spectrometry — Flame. Electrothermal. Cold vapor.
Atomic fluorescence spectrometry
Optical emission spectrometry — Flame. DC arc. Spark. ICP. Microwave plasma.
X-ray spectrometry — Fluorescence. Electron probe. Particle-induced.
Mass spectrometry — Spark source. Ion probe. Isotope dilution.
Activation analysis — Neutron. Charged particle. Photon.
Isotope dilution-substoichiometric method
Polarography — DC. AC. Linear sweep. Square-wave. Pulse.
Stripping voltammetry — Anodic. Cathodic.
Electrochemical titrimetric methods
Ion-selective electrode potentiometry
Gas analytical methods
Gas chromatography
Liquid chromatography
Catalytic methods

---

(4) when the desired trace elements are not homogeneously distributed in the sample,

(5) when suitable standard samples required for the calibration are not available, and

(6) when the chemical or physical states of the sample are not suitable for the direct determination.

Decomposition (destruction) of the sample [2—4] and/or enrichment (preconcentration) of the desired trace elements prior to the determination can overcome these difficulties, extend the absolute and relative detection limits, improve the precision and accuracy of analytical results, and widen the scope of the determination techniques. The outlines of these two steps are tabulated in Tables 2 and 3. Sometimes, both steps are unified, e.g. in dry oxidation (ashing) of organic samples.

Table 2. Decomposition techniques used in inorganic trace analysis

---

For inorganic solids
Dissolution in mineral acids, organic acids, alkali hydroxide solutions, etc.
Fusion or sintering with alkali carbonates and hydroxides, sodium peroxide, alkali disulfates, etc.
Decomposition with reactive gases such as oxygen, chlorine and hydrogen fluoride.
Anodic dissolution (electrolytic dissolution)

For organic samples
Dry oxidation (dry ashing) in air, oxygen or oxygen plasmas.
Wet oxidation (wet ashing) with mineral acids.
Oxidative fusion with alkali nitrates.

Table 3. Enrichment techniques used in inorganic trace analysis

Sample state	Separated substances	Techniques
1. Solid or molten	Particles*	Manual selection under the microscope. Sieving. Magnetic separation. Heavy liquid separation. Flotation.
	Constituents	Selective dissolution.
		Electrolytic dissolution. Sublimation. Extraction of gases in metals at high temperatures. Dry oxidation of organic samples. Zone melting. Fire assay.
2. Solution	Particles*	Filtration. Centrifugation. Flotation.
	Solutes	Precipitation. Electrodeposition. Adsorption. Molecular sieving. Ion exchange. Liquid-liquid extraction. Volatilization. Flotation. Freezing. Electrophoresis. Dialysis. Ultrafiltration. Ultracentrifugation.
3. Gas	Particles*	Filtration. Impaction. Sedimentation. Centrifugation. Thermal precipitation. Electrostatic precipitation.
	Constituents	Absorption. Adsorption. Condensation. Permeation.

* Particle diameter > ca. 0.5 μm

Enrichment is a process in which the ratio of the amount of a desired trace element to that of the original matrix is increased. Foreign substances are frequently introduced into the sample during the enrichment process. Thus the original matrix is converted into a new matrix which is more suitable for the determination; e.g. a 1-g solid metal sample into a 10-ml aqueous solution containing the desired trace element and a microgram quantity of the original matrix. Therefore, enrichment does not necessarily mean increase in the concentration of the desired trace element. Enrichment is attained by the use of various separation techniques based on physical, physicochemical and chemical principles. Most of frequently used enrichment techniques including liquid-liquid extraction and ion exchange depend on distribution of the elements of interest between two phases followed by physical separation of both phases.

2 General Aspects of Enrichment Techniques

Important considerations in selecting and evaluating enrichment techniques are discussed in this chapter:
trace recovery,
enrichment factor,
contamination,
simplicity and rapidity,
and sample size.

2.1 Trace Recovery

The trace recovery (R_T) or the yield of the desired trace element is defined as

$$R_T = Q_T/Q_T^0 \times 100 \ (\%) \tag{1}$$

where Q_T^0 and Q_T are the quantities of the desired trace element before and after the enrichment, respectively, the latter being corrected for contamination.

The trace recovery is usually less than 100%, because loss of the desired trace element may occur during the decomposition and enrichment steps by evaporation, incomplete decomposition, incomplete separation, careless manipulation, and strong adsorption on the walls of the containers and other apparatus used. The trace recovery varies with concentration levels. In general, the lower the concentration, the more the danger of losses. The well-known anomalous behavior of trace elements at extremely low concentrations [5, 6] is frequently responsible for the loss. Trace recoveries of greater than 95%, or 90% at least, are required in most inorganic trace analyses. If sufficiently reproducible, lower trace recoveries can be used for the correction of analytical results. Much lower and even variable recoveries are permissible in isotope dilution analysis and radiochemical separations using isotopic carriers.

The recovery and loss of the trace element are best investigated by the *radioactive tracer technique*. A radioactive isotope of the trace element is added as tracer to the sample before the enrichment step, and its behavior is followed by rapid, sensitive and selective radioactivity measurements. The great advantage of this technique is that the recovery and loss are measured independently of contamination hazards. Although both the isotope and the radiation effects are generally negligible, it must be kept in mind that the radioactive isotope added is in the same chemical form as the desired trace element. A limitation of the applicability of this technique exists in the difficulty of introducing tracers into solid samples for investigating the recovery and loss of trace elements during decomposition of solid samples as well as separations by volatilization and solvent extraction of trace elements from solid samples. Radio-activation (irradiation of solid samples with thermal neutrons) or synthesis of solid samples containing radioactive isotopes is sometimes useful.

When suitable radioactive isotopes are not available, *standard samples*, i.e. certi-fied standards, analyzed samples or synthetic samples, are used for measuring trace re-coveries. The method of standard addition is also useful. In all these cases, however, contamination should be negligible or reproducible and accurately determinable.

2.2 Enrichment Factor

The enrichment factor (F) or preconcentration coefficient of the trace element is de-fined as

$$F = \frac{Q_T/Q_M}{Q_T^0/Q_M^0} = \frac{R_T}{R_M} \tag{2}$$

where Q_M^0 and Q_M are the quantities of the matrix before and after the enrichment, respectively, and R_M is the yield of the matrix. The enrichment factor required de-pends on the concentration level of the desired trace element in the sample as well as on the determination techniques used. Enrichment factors greater than 10^5 are some-times required, which can be easily attained by some enrichment techniques with satisfactory trace recoveries. In most inorganic trace analyses, however, enrichment factors of 10^2 to 10^4 are sufficient, because modern instrumental determination tech-niques have low detection limits and adequate selectivity. Enrichment factors can be increased by using proper multistage separations without appreciable loss of the desired trace element. The ratio of the concentration of the desired trace element after the enrichment to that of the original sample is virtually the same as the enrich-ment factor when the matrix does not change, but is not when the matrix conversion occurs during the enrichment.

2.3 Contamination

During the enrichment and related steps, contaminants containing the desired trace elements may be introduced into the sample from external sources including
the laboratory atmosphere,
the reagents,
containers and other apparatus used,
and the analyst performing the analysis,
which give positive biases in analytical results. Airborne dust particles can adsorb the desired trace elements in the sample solution and cause negative biases. Some extern-

ally introduced foreign inorganic or organic substances may interfere with the determination and cause positive or negative biases in analytical results.

In inorganic trace analysis, a "blank run" is usually carried out in parallel with the analysis under the same conditions but without the sample, and the resulting "blank value" is subtracted from the analytical value. This method, however, is quite unsatisfactory to correct the effects of contamination because of the following three reasons:

(1) Most kinds of contamination are not reproducible. For example, the degree of airborne contamination varies with time and place. The contamination due to surface erosion of containers depends greatly on the history of the containers, the cleaning procedures used, as well as the solution composition in which the analysis and the blank run differ.

(2) When the standard deviation of the analytical value A is σ_A and that of the blank value B is σ_B, the relative standard deviation of the corrected value $(A - B)$ is $100 \sqrt{\sigma_A^2 + \sigma_B^2}/(A - B)$ %. When $A \gg B$ and $\sigma_A \gg \sigma_B$, the relative standard deviation becomes $100\sigma_A/A$%, which is usually sufficiently small. As the difference between A and B decreases, the relative standard deviation increases, and when $A \approx B$ and $\sigma_A \approx \sigma_B$, it is $141\sigma_A/(A - B)$%, which is quite large. Therefore, even when the contamination is reproducible, the corrected value is inprecise for an analytical value having a corresponding blank value of the same order of magnitude.

(3) Loss of the desired trace elements may occur simultaneously with contamination during the analysis and the blank run in different manners. Accidentally, loss and contamination my cancel each other and an uncorrected analytical value with an apparent trace recovery of about 100% or a blank value of nearly zero may be obtained, although appreciable loss and contamination exist.

The degree of contamination is estimated more reliably by analyzing standard samples or by carrying out the analysis with various sample weights and extrapolating the analytical values to the sample weight zero. These methods, however, are still unsatisfactory.

Therefore, minimization of contamination is essential to attain accurate analytical results in inorganic trace analyses. If possible, the degree of contamination should be less than one tenth of an analytical value. It is desirable to investigate loss (by the radioactive tracer technique) and contamination in each analytical step separately, and correction is made according to these results, instead of simple subtraction of the overall blank value from the analytical value. Differentiation of contamination from various sources is possible under proper experimental conditions: e.g. contamination due to a reagent is estimated separately by analyzing a large amount of the reagent under the conditions where contamination due to the atmosphere and containers is negligible.

2.4 Simplicity and Rapidity

Most analysts may think enrichment techniques take much skill, labor and time. Therefore, enrichment techniques should be as simple, easy and rapid in operation as possible. Efforts to this goal are often also effective to minimize loss and contamination. It must be kept in mind that sometimes combination of two or more simple enrichment techniques is superior to a single difficult enrichment technique as a whole. Simultaneous multielement enrichment techniques are preferable for simultaneous multielement determination techniques such as optical emission spectrometry and X-ray fluorescence spectrometry and also for rapid single-element (one-element-at-a-time) determination techniques such as atomic absorption spectrometry. Smooth connections with the preceding steps (sample preparation or decomposition) and the following steps (another enrichment or determination) are very important. Some enrichment techniques are unified with another analytical step; e.g. enrichment of particulate matter in air and natural water samples (with sample preparation), oxidation of organic samples and fire assay of ores (with decomposition), and carrier distillation in optical emission spectrometry (with determination).

2.5 Sample Size

The required sample size depends on the concentration levels of the desired trace elements as well as the absolute detection limits of the determination technique used. Ordinarily, solid samples of 0.1 to 10 g and liquid samples of 10 to 1000 ml are taken for the enrichment of trace elements at the ng/g or low μg/g level. Much larger samples are sometimes used for trace elements at the pg/g or low ng/g level. Theoretically, it would be possible to determine infinitely low concentrations of trace elements by applying enrichment techniques to an infinitely large sample. In reality, *the lowest detectable concentration levels are limited by contamination, loss, and interferences from other trace constituents* present in the sample, and increasing the sample size becomes useless for extending the relative detection limit. In addition, manipulations for decomposition and enrichment become very difficult and inconveniently time-consuming when the sample size is extremely large.

Some kinds of samples, such as ultrahigh-purity metals and compounds, and other rare natural and artificial substances, are very expensive or available only in small quantities. Modern instrumental determination techniques with excellent absolute detection limits enable one to determine trace elements at the ng/g or low μg/g level in milligram samples. Enrichment techniques operated on a microscale (microliter levels) are frequently very useful in the effective application of such determination techniques as optical emission spectrometry, spark source mass spectrometry, atomic absorption and fluorescence spectrometry by electrothermal atomization or by flow-injection techniques, and electron and ion microprobe techniques, where the maximum size of a solid or liquid sample is limited to the low microliter level or lower. Microscale operations have several advantages, i.e. economy in sample, high-purity reagents and time, as well as minimization of experimental wastes. Microscale operations, however, sometimes need more skill and caution than conventional ones to ensure the precision and accuracy of analytical results.

3 Control of Contamination and Loss

As discussed in the preceding chapter, contamination and loss are among the most difficult problems in inorganic trace analysis, especially when enrichment techniques are employed. Their control is essential to obtain precise and accurate analytical results.

3.1 Airborne Contamination

The outdoor atmosphere contains various kinds of liquid and solid particulates, aerosols and dusts, which include soil dusts, sea salt nuclei, volcanic ash, pollens, bacteria and others of natural origin, as well as fly ash, oil smoke, sulfuric acid mist, cement dust and others of industrial or man-made origin. These particulates intrude into conventional analytical laboratories. Fig. 2 shows approximate sizes of such particulates [7]. Table 4 tabulates examples of trace elements in particulates in the atmosphere of various districts.

Table 4. Trace elements in airborne particulates

Concentration level (ng/m³ air)	East Chicago, Ind., U.S.A. (industrial) [8]	Niles, Mich., U.S.A. (rural) [8]	U.S.A. (urban, average) [9]	Osaka and Sakai, Japan (urban) [10]	South pole [11]
$10^4 - 10^5$	Fe, S	S			
$10^3 - 10^4$	Al, Ca, Cu, K, Mg, Zn	Al, Ca, Fe	Fe	Al, Ca, Cl, Fe, K, Na, Zn	
$10^2 - 10^3$	Cr, Mn, Na, Ti	Cu, K, Mg, Na, Ti, Zn	Mn, Pb, Zn	Mn, V	
$10 - 10^2$	As, Br, Ce, Sb, V	Br, Mn	As, Cr, Cu, Ni, Sn, Ti, V	As, Ba, Br, Cd, Cr, Ni, Sb, Ti	
$1 - 10$	Ag, Co, Ga, Hg, La, Sc, Se, Th, W	As, Cr, Hg, La, Sb, Sc, Se, V	Cd, Sb	Ag, Ce, Co, Cs, Hg, La, Rb, Se, Th, W	Mg, Na

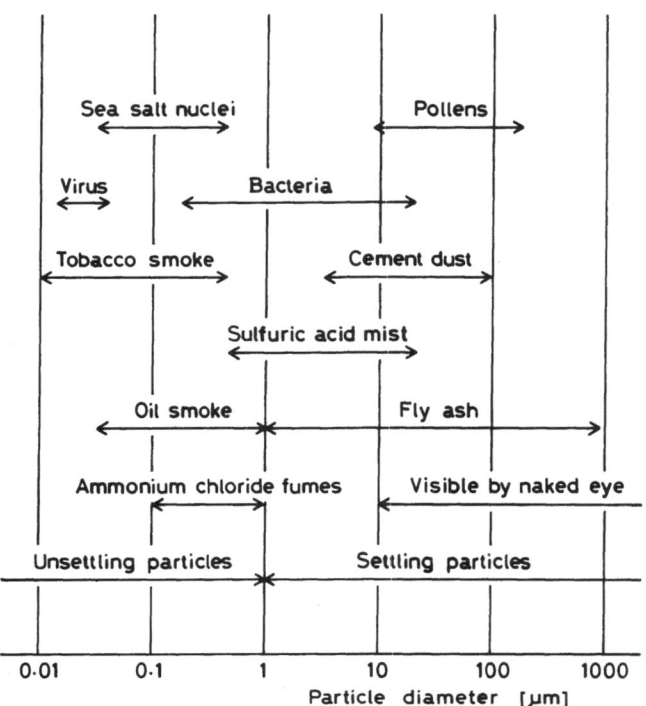

Fig. 2. Approximate sizes of airborne particulates

Airborne particulates also originate from fixtures, furniture, apparatus, reagents and analysts (hair, skin, garments and cosmetic) in laboratories. For example, metal piping and gas cylinders may become rust-particle emitters. Dusts in laboratories act as carriers of various other contaminants.

Gaseous contaminants such as ammonia, hydrogen chloride and mercury vapor often exist in the laboratory atmosphere.

The kinds and quantities of airborne contaminants differ from laboratory to laboratory and also from time to time. According to the location, construction, history and operational conditions of the laboratory, *almost all elements can become airborne contaminants,* which cause contamination at nanogram, or sometimes even microgram, levels during the analytical manipulation carried out under open conditions.

To minimize airborne contamination, clean rooms, clean hoods or benches and various closed systems are effectively employed.

3.1.1 Clean Rooms

Use of a dust-free clean room [7, 12] is most desirable, though it is expensive to construct and maintain. It was reported [13] that the atmosphere in a clean room contained 1 ng Fe, 2 ng Cu and 0.2 ng Pb per m^3, whereas that in a conventional chemical laboratory contained 200 ng Fe, 20 ng Cu and 400 ng Pb per m³. The requirements for design and use of clean rooms are as follows:

(1) Surface materials used for floor, walls, ceiling, fixtures and furniture should be non-porous and resistant to oxidation, corrosion, abrasion, chipping and flaking. Other considerations include low dust particle adhesion and easy cleaning.

(2) The interior should be as simple as possible to prevent the accumulation of dust and facilitate cleaning. From the same reason, seamless structure and rounded corners of the rooms are preferable.

(3) The pressure of the interior should be kept a little higher than that of the out-side to prevent the intrusion of the outside atmosphere.

(4) The air filtered through high-efficiency particulate air (HEPA) filters should be continuously supplied to the interior. The HEPA filters, which were first developed to remove fissionable particulates for the Manhattan Project, are composed of thin porous sheets of ultrafine fibers (< 1 μm in diameter), 100% glass fiber, or a combination of glass and asbestos fibers. Standard HEPA filters have a minimum efficiency rating of 99.97% for 0.3-μm particles. The HEPA filters having a 99.99% efficiency for 0.1-μm particles are now commercially available. Fig. 3 shows three types of clean rooms classified by airflow patterns. The airflow rate is about 20 to 50 cm/s at the filter surface. Most of the air is circulated and some fresh air is supplied if necessary. The class 100 cleanliness level specified in the U.S. Federal Standard 209a, i.e. less than 100 particles per cubic foot (1 cubic foot ≈ 28 liters) of air larger than 0.5 μm and no particle larger than 5 μm, is attained over the whole work area in the most

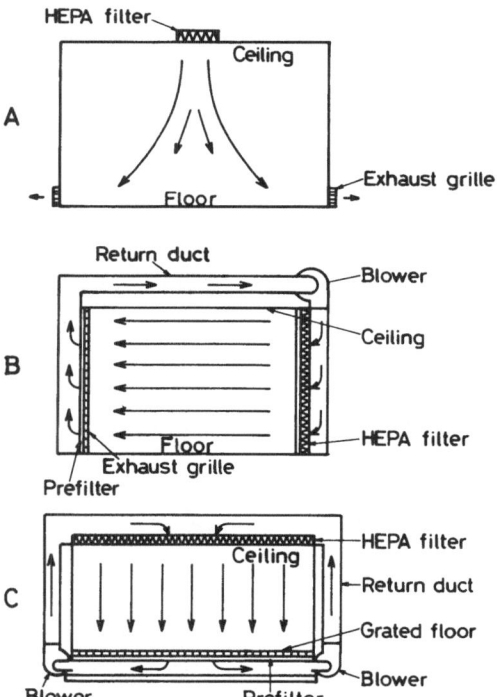

Fig. 3. Clean rooms. A: Nonlaminar-airflow,
B: Horizontal-laminar-airflow,
C: Vertical-laminar-airflow

expensive vertical-laminer-airflow room. In the class 100 cleanliness, the greatest number of particles are below 0.5 μm in diameter, but the weight of the particles in this range is less than 2% of the total weight of the particles. The class 100 cleanliness is generally limited to the narrow upstream area in the horizontal-laminar-airflow room. The cleanliness is much lower in the least expensive nonlaminar-airflow room.

(5) Gaseous contaminants should be removed by activated carbon filters.

(6) The relative humidity should be kept at about 50%. Lower humidity can develop electrostatic charges and cause serious problems in particle attraction on apparatus or with explosive compounds, whereas higher humidity may rust metallic parts.

(7) The room should be always kept clean.

(8) The analyst should wear proper coats, hats, shoe covers and gloves, and apply air shower and adhesive mats before entering the clean room.

(9) Meaningless motion of the analyst should be avoided to minimize the generation of dust.

3.1.2 Clean Hoods or Benches

Laminar-airflow clean hoods or benches [7, 12] are very useful. The class 100 cleanliness is achieved in the hoods installed in conventional chemical laboratories. Some types are shown in Fig. 4.

3.1.3 Other Means to Reduce Airborne Contamination

Use of glove boxes, evaporation chambers (Fig. 5) and various closed systems is effective to reduce airborne contamination by a factor of 10 or more in conventional chemical laboratories. Analytical operations under open conditions should be as short as possible, and closed vessels must always be used for storage even in clean rooms or hoods.

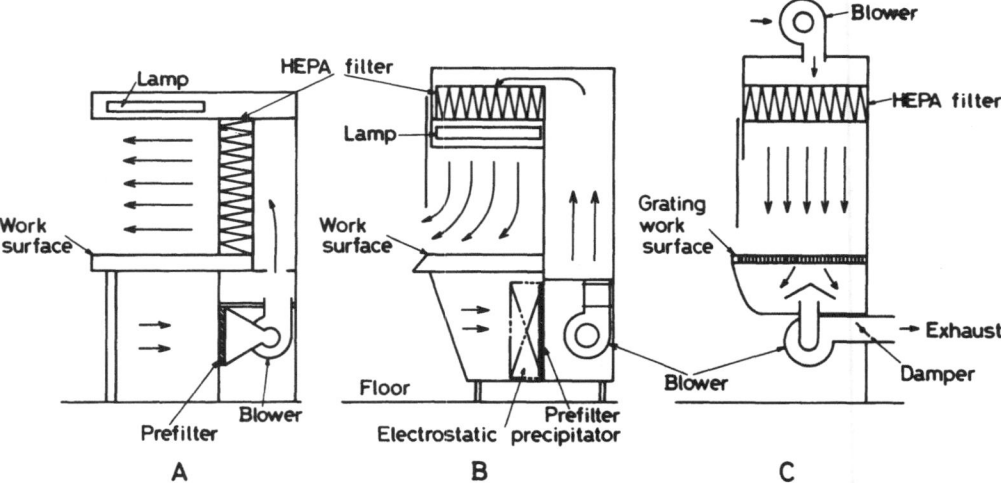

Fig. 4. Clean hoods or benches. A: Horizontal-laminar-airflow, B: Vertical-laminar-airflow, C: Balanced-laminar-airflow

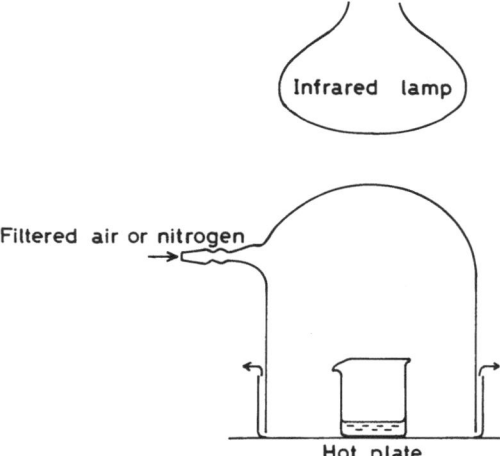

Fig. 5. Borosilicate glass evaporation chamber

3.2 Contamination and Loss Due to Apparatus

The surfaces of containers and other apparatus which come into direct contact with the samples may be attacked to some extent, resulting in contamination. On the other hand, various elements are often strongly adsorbed on the container walls [14, 15] and the desorption is very slow or difficult; this causes not only loss of the desired trace element in the present sample but also contamination of later samples treated in the same container. The extent of the above two phenomena depends greatly on container material, trace element concentration, solution composition, temperature and time. Thirdly, permeability of the container walls may cause loss and contamination. These phenomena become critical in inorganic trace analysis, especially in prolonged storage of solution samples and reagents, comminution of hard solid samples, decomposition of samples, etc.

3.2.1 Selection of Materials

Table 5 shows container materials commonly used in inorganic trace analysis. Considerations in the selection of the materials include chemical composition, chemical resistance, thermal stability, permeability, adsorption and desorption behavior, and price. There exist various kinds of undesirable trace impurities in *commercial inorganic and plastic materials* (Tables 6 and 7). It must be kept in mind that the impurity concentrations vary greatly with manufacturers, grades, and even lot numbers. For example, no impurity element is found at the $\mu g/g$ level in ultrahigh-purity vitreous silica [20].

Plastic materials are generally recommended in inorganic trace analysis. After 16 months' storage of 0.1 M sodium hydroxide solution at 24 °C, contamination of 1.5 $\mu g/g$ each of aluminum and boron was observed in a Pyrex container, whereas no contamination was detected in a polyethylene container [21, 22]. Contamination

Table 5. Container materials commonly used in inorganic trace analysis

Material	Maximum service temperature ($^{\circ}$C)	Poor chemical resistance [16] to	Permeability
Pyrex* (borosilicate glass)	600	Hydrofluoric acid, Conc. phosphoric acid, Sodium hydroxide solution	None
Vycor** (high-silica glass)	900	Hydrofluoric acid, Conc. phosphoric acid, Sodium hydroxide solution	None
Vitreous silica	1100	Hydrofluoric acid, Conc. phosphoric acid, Sodium hydroxide solution	None
Platinum	1500	Aqua regia	None
Glassy carbon [17]	600	None	None
Polyethylene	80 (High-pressure process) 110 (Low-pressure process)	Organic solvents, Conc. nitric acid, Conc. sulfuric acid	Permeable
Polypropylene	130	Organic solvents, Conc. nitric acid, Conc. phosphoric acid, Sodium hydroxide solution	Permeable
Teflon (polyfluorocarbon)	250	None	Permeable

* SiO_2 81, B_2O_3 13, Na_2O 4, K_2O 0.5, Al_2O_3 2 (wt%)
** SiO_2 96, B_2O_3 3, Al_2O_3 0.5 (wt%)

Table 6. Impurities in inorganic container materials – examples

Concentration level (μg/g)	Pyrex [18]	Vycor [18]	Vitreous silica [18]	Platinum [12]	Glassy carbon [17]
$10^2 - 10^3$	Ca, Cl, Fe, Mg, Zr	As, Fe, Mg, Na, Ti, Zr			
$10 - 10^2$	F, Ga, Hf, Li, Mn, Ni, S, Sr, Ti, V	Ca, Cl, Hf, K, Zn	Ca, Fe, Mg, Zn	Au, Pd	Ca, Si, Sn, Ti
$1 - 10$	As, Ba, Bi, Cr, Cu, P, Pb, Sb, Se, Y, Zn	Ag, Cu, F, Ga, Li, Mn, Ni, S, Sb, Sn	Ag, Al, Cl, Cu, F, Ga, Hf, K, Li, Na, Ni, Ti, Zr	Ag, Cu, Fe, Rh, Ti	Al, Fe, V

Table 7. Impurities in plastic container materials − examples [19]

Concentration level (μg/g)	Polyethylene (high-pressure process)	Polyethylene (low-pressure process)	Teflon
$10-10^2$	Ca		Na
$1-10$	Al, Fe, Na, Si	Al, Na	
$10^{-1}-1$	Cr, Mg, Pb, Sr	Co, Cr, Zn	
$10^{-2}-10^{-1}$	Ag, B, Ba, Zn		Cu, Fe
$10^{-3}-10^{-2}$	Co, Cu		Co, Zn

Table 8. Emission spectrographic analysis of mineral acids after evaporation in various dishes [23]

Acid	Dish material	Impurities determined (ng/g)										
		Al	Fe	Ca	Cu	Mg	Mn	Ni	Pb	Ti	Cr	Sn
HF	Polyfluorocarbon	3	3	1	<0.04	< 3	0.1	<0.4	<0.1	0.1	<0.4	ND
	Platinum	10	10	10	0.4	10	0.2	0.8	0.5	1	0.5	ND
HCl	Polyfluorocarbon	< 4	3	5	0.2	3	0.1	ND	<0.4	ND	ND	ND
	Platinum	2	2	10	1	6	0.2	0.6	<0.4	0.4	Tr	<0.4
	Vitreous silica	10	10	60	1	10	0.4	2	0.5	2	0.6	0.4
HNO$_3$	Polyfluorocarbon	2	8	4	≤0.01	7	0.1	ND	ND	Tr	ND	ND
	Platinum	20	20	30	0.4	20	0.6	Tr	1	0.8	ND	ND
	Vitreous silica	20	20	60	0.1	20	0.6	ND	1	0.3	ND	ND

ND: not detected. Tr: traces

Table 9. Contamination during comminution* [27]

Grinding surface and device	Hardness on Mohs' scale	Contamination (μg/g)
Tungsten carbide (WC 96%): vial and ball	8−9	Ti 124, Co 32, Cu 3, Zn 2
Boron carbide (B$_4$C): mortar and pestle	9	B 23, Cu 3, Zn 2
Alumina (sapphire): mortar and pestle	9	Al > 2000, Cr 225, Fe 9, Ga 3, B 2, Mn 1
Alumina ceramic (Al$_2$O$_3$ 96%): vial and ball	9	Al > 2000, Fe 34, Ga 21, Ti 11, Cu 3, Zn 3, Li 1
Silica (agate): mortar and pestle	6−7	B 2, Cu 1

* Grinding of high-purity SiO$_2$ powder from >100 mesh to <200 mesh

during the evaporation of mineral acids in polyfluorocarbon dishes was less than that in platinum or vitreous-silica dishes (Table 8) [23]. On the other hand, organic contamination may occur in plastic containers. Contamination of water stored in polyethylene bottles with phthalic acid ester plasticizers was reported [24]. Another problem is the permeability of plastic materials. For example, the mercury contamination of water samples stored in polyethylene containers may be caused by passage of mercury vapor from the ambient air through the container wall into the solution [25, 26]. Therefore, storage in glass bottles is recommended in this case.

For materials used for mortars, pestles and other comminution devices, hardness is an additional consideration. Table 9 shows typical devices and examples of contamination during comminution [27]. Contamination from metal sieves also is not negligible. To prevent contamination, a rock sample is ground with a mortar and pestle made of the same rock [28]. Acrylic resin grinding vials and nylon sieves are sometimes useful.

Materials used in filtration devices (e.g. paper, membrane and glass filters) may cause contamination as well as adsorption loss of trace elements [19].

3.2.2 Surface Treatment

To minimize contamination and loss due to containers, coating or formation on the surface of thin films having superior properties is sometimes effective. Pyrex dishes coated with 78 SiO_2-21 ZrO_2-1 Na_2O (mol%) films about 50 nm thick are successfully used in the dry oxidation of botanical samples for boron determination and in the evaporation needed for aluminum determination in nitric acid [29]. Pretreatment of Pyrex beakers with a molten calcium nitrate and potassium nitrate (20 : 80 mol%) mixture is effective to prevent the sodium contamination in wet oxidation of botanical samples [30]. Diffusion loss of silver on Pyrex surfaces, which occurs in the decomposition of organic samples and evaporation of solutions containing silver, can be minimized by pretreatment of the surfaces with molten potassium nitrate [31].

3.2.3 Cleaning of Containers

All containers should be carefully cleaned before use according to their material, history and use.

Cleaning agents for glass ware include mineral acids and their mixtures, acetone, methanol, ethanol and detergents. Classic chromic acid cleaning solution is still popular because of its cleaning power, but it must be kept in mind that appreciable amounts of chromium tenaciously remain on glass surfaces even after careful rinsing with water. Although rinsing with aqueous ammonia and chelating agents is effective to remove the remaining chromium, the use of a one-to-one mixture of concentrated sulfuric acid and nitric acid is recommended instead of chromic acid cleaning solution [22]. Glass ware can also be effectively cleaned with a dilute solution of hydrofluoric acid and its mixtures with other mineral acids [18—20]. This procedure, however, may produce freshly roughened surfaces from which impurities can be more readily leached. Any cleaning procedure should be followed by copious rinsing with *purified* water. Elevated temperatures, vapor treatment [32], and ultrasonic irradiation accelerate the cleaning.

The same cleaning agents can be used for plastic ware [33–35]. Numerous fine particles (0.5 mm or smaller) containing Al, Ca, Cr, Cu, Fe, Mg, Mn, Ni, Si, Ti and Zn, which may be introduced into plastic ware during the fabrication process, are effectively removed by vigorous prolonged cleaning with hot hydrochloric acid and nitric acid or their mixtures [36].

Platinum ware is cleaned by sodium disulfate fusion followed by soaking in hydrochloric or nitric acid.

To avoid airborne contamination, containers should be cleaned immediately before use or stored under cover after cleaning.

3.3 Contamination Due to Reagents

Impurities present in the reagents used in relatively large amounts for the sample decomposition and separations become serious sources of contamination. This kind of contamination is generally reproducible, except when solid reagents containing heterogeneously distributed impurities are directly used. Reagents often contain tiny dust particles, which may cause adsorption loss of trace elements in solutions.

3.3.1 Selection of Commercial Reagents

Various high-purity reagents for chemical analysis and semiconductor industry are now commercially available and can be satisfactorily applied in inorganic trace analysis. However, they still contain appreciable amounts of impurities, which may not be tolerable in certain cases. Selection of available commercial reagents is important, because kinds and quantities of impurities in them differ greatly with grades, manufacturers, lot numbers, and even bottles. It must be kept in mind that higher grade reagents may contain a larger amount of a certain impurity introduced by contamination during the purification process. Another consideration is contamination due to containers during storage. Glass containers with inverted ground stoppers (a male joint attached directly to a container and a female joint cap) are preferable to minimize contamination during transfer operation. *Plastic containers* fitted with tight sealing caps are also used satisfactorily for storage. The outside of the containers should be carefully cleaned before use.

3.3.2 Preparation of High-Purity Reagents in Analytical Laboratories

When commercial high-purity reagents are unsatisfactory for use, too expensive, or not easily obtainable, analysts themselves must prepare the reagents of required purity. Purification or preparation of high-purity reagents should be carefully carried out using proper apparatus in a clean environment. Selected laboratory methods are summarized in Table 10, further details being given as follows.

(1) *Distillation.* This method is widely used for the purification of water, mineral acids, aqueous ammonia and organic solvents. Conventional or boiling distillation is rapid, but the purity of the distillate may be limited by entrainment of liquid particulates in the vapor stream formed during bubble rupture and by creeping of the un-

Table 10. Selected laboratory methods for preparing high-purity reagents

Reagents	Methods
Water	Distillation. Ion exchange
Hydrochloric acid	Distillation. Isothermal distillation. Dissolution of hydrogen chloride in water. Ion exchange
Hydrofluoric acid	Distillation. Isothermal distillation. Dissolution of hydrogen fluoride in water
Hydrobromic acid	Dissolution of hydrogen bromide in water. Ion exchange
Nitric acid	Distillation
Perchloric acid	Distillation
Sulfuric acid	Distillation
Phosphoric acid	Dissolution of phosphorus pentoxide (purified by sublimation) in water
Aqueous ammonia	Distillation. Isothermal distillation. Dissolution of ammonia in water
Sodium or potassium hydroxide solution	Conversion of sodium or potassium chloride (purified by extraction) with an OH-form anion exchanger. Liquid-liquid extraction
Salts of alkali and alkaline earth elements	Filtration. Recrystallization. Coprecipitation. Electrolysis. Ion exchange. Liquid-liquid extraction. Zone melting. Acid plus base
Organic solvents	Distillation. Back-extraction

Table 11. Impurities in water and mineral acids purified by sub-boiling distillation — examples [37]

Impurity elements (ng/g)	Water	Hydrochloric acid (31 wt%)	Nitric acid (70 wt%)	Perchloric acid (70 wt%)	Sulfuric acid (96 wt%)	Hydrofluoric acid (48 wt%)
Pb	0.008	0.07	0.02	0.2	0.6	0.05
Tl	0.01	0.01	–	0.1	0.1	0.1
Ba	0.01	0.04	0.01	0.1	0.3	0.1
Te	0.004	0.01	0.01	0.05	0.1	0.05
Sn	0.02	0.05	0.01	0.3	0.2	0.05
In	–	0.01	0.01	–	–	–
Cd	0.005	0.02	0.01	0.05	0.3	0.03
Ag	0.002	0.03	0.1	0.1	0.3	0.05
Sr	0.002	0.01	0.01	0.02	0.3	0.1
Se	–	–	0.09	–	–	–
Zn	0.04	0.2	0.04	0.1	0.5	0.2
Cu	0.01	0.1	0.04	0.1	0.2	0.2
Ni	0.02	0.2	0.05	0.5	0.2	0.3
Fe	0.05	3	0.3	2	7	0.6
Cr	0.02	0.3	0.05	9	0.2	5
Ca	0.08	0.06	0.2	0.2	2	5
K	0.09	0.5	0.2	0.6	4	1
Mg	0.09	0.6	0.1	0.2	2	2
Na	0.06	1	1	2	9	2
Total	0.5	6.2	2.3	16	27	17

rectified liquid. To overcome these difficulties, sub-boiling distillation units made of vitreous silica [32, 37], Teflon [37—39] and polypropylene [32, 40] are effectively used (Fig. 6). The *infrared radiator* vaporizes the surface without boiling the liquid. Production rates (ml/24 h) are, for example, 4000 for water, 2000 for hydrochloric acid, and 300 to 600 for other mineral acids. Table 11 shows examples of high-purity water and mineral acids prepared by this method [37]. Organic impurities in water are removed by oxidative distillation from alkaline potassium permanganate solutions or more satisfactorily by catalytic pyrodistillation [41].

(2) *Isothermal or isopiestic distillation* [42—44]. This technique is useful for the preparation of small quantities of high-purity volatile acids and aqueous ammonia. The reagent-grade reagent and high-purity water are placed separately in a closed vessel such as a desiccator (the reagent in the lower compartment and the water in a vessel in the upper compartment), and the vapor is allowed to diffuse into the water at room temperatures under isothermal conditions. Thus, high-purity 10 M hydrochloric acid was obtained from 500 ml of concentrated hydrochloric acid (ρ = 1.18) and 50 ml of water after 3 days, and high-purity 9.5 M aqueous ammonia from 500 ml of aqueous ammonia (ρ = 0.880) and 50 ml of water after 2 days.

(3) *Dissolution of gases in water.* Cylinder gases are purified by filtration and scrubbing and then absorbed in high-purity water. Concentrated sulfuric acid and sodium fluoride suspension are used for scrubbing hydrogen chloride and hydrogen fluoride, respectively.

(4) *Ion exchange.* This technique is widely used for purifying water. Although ions are removed almost perfectly (e.g. less than 0.5 ng/g of heavy metal ions), non-ionic species including colloids, particulates and organic compounds are not. Another problem is organic contaminants such as nitrogen compounds originating from soluble constituents of ion exchange resins. Heavy metal impurities in hydrochloric acid are removed by percolating through a Cl-form anion exchange resin column. Ion exchange is also useful to convert salts into corresponding acids or bases; e.g. from sodium or potassium bromide to hydrobromic acid with an H-form cation exchange resin, and

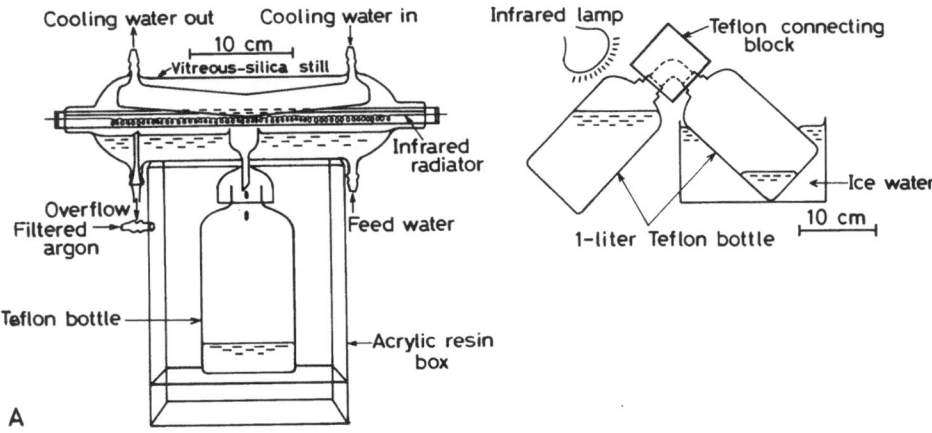

Fig. 6. Sub-boiling distillation apparatus. A: Vitreous silica, B: Teflon (After Mattinson [38])

from sodium or potassium chloride to sodium or potassium hydroxide with an OH-form anion exchange resin. Similarly, this technique can be used in conversion of salts; e.g. from sodium salts to corresponding potassium salts, and from sulfates to corresponding chlorides.

(5) *Filtration.* Impurities such as iron and copper in solid reagents often exist as particulate matter. A few $\mu g/g$ of iron in sodium carbonate and calcium nitrate was reduced by two orders of magnitude by simple filtration of the solutions through a membrane filter [12].

(6) *Recrystallization.* This technique is effective for the purification of solid reagents, except for soluble impurities concentrated by mixed-crystal formation.

(7) *Coprecipitation* [12]. Heavy metal impurities in aqueous solutions of alkali and alkaline earth salts are removed by coprecipitation with aluminum of lanthanum hydroxide or with indium sulfide. Traces of lead are coprecipitated with barium sulfate.

(8) *Electrochemical deposition.* Heavy metal impurities in aqueous solutions of alkali and alkaline earth salts are removed by mercury cathode electrolysis. Silver at the ng/g level in 2-mercaptobenzothiazole is reduced by more than one order of magnitude by deposition of the silver on mercury globules in an acetone solution [45].

(9) *Liquid-liquid extraction.* Heavy metal impurities in neutral solutions of alkali and alkaline earth salts are extracted with a chloroform solution of oxine or a carbon tetrachloride solution of dithizone. Iron and other heavy metal impurities in 10 M sodium hydroxide are removed by precipitation and extraction with phenyl-2-pyridyl ketoxime [46]. Organic solvents are purified by back-extraction with acids, bases and salt solutions.

3.4 Other Sources of Contamination and Loss

The analyst may become serious sources of contamination. Touching the surfaces of solid samples or apparatus, which will come into direct or indirect contact with samples, with fingers can cause contamination with such elements as Ca, Cl, Na and Pb, as well as organic matter. Thus *the use of plastic gloves is essential,* e.g. in the determination of traces of chlorine. Other sources of contamination include careless manipulation of the analyst and other work being carried out in the same room (cross-contamination).

In addition to adsorption loss on container walls, loss of the desired trace elements may occur from

(1) partial evaporation of the desired trace elements and formation of small amounts of insoluble residues containing the desired trace elements during the sample decomposition

(2) anomalous colloidal behavior of trace elements in extremely dilute solutions

(3) intrinsic imperfection in enrichment techniques and

(4) careless manipulation of the analyst.

Contamination and loss specific to each enrichment technique will be discussed in the corresponding chapter. The proceedings of the 7th Materials Research Symposium held at the U.S. National Bureau of Standards in October, 1974, contain useful information on contamination and loss in inorganic trace analysis [47].

4 Volatilization

Volatilization is a process in which gaseous and volatile constituents are transferred to the gas phase from liquid or solid samples. Thus it includes evaporation, distillation, gas evolution and sublimation. Generally, either the desired trace elements or the matrix can be selectively volatilized at temperatures where the volatility or vapor pressure of one is sufficiently large and that of the other is negligibly small. Therefore, the constituents in the sample are frequently converted into more favorable compounds before or during the volatilization step. Table 12 lists some useful volatile compounds used in this enrichment method.

The operational temperature varies greatly with techniques. In lyophilization (freeze-drying), water is sublimed from frozen samples. In isothermal distillation, volatile compounds such as ammonia spontaneously diffuse in a closed cell (Fig. 7) from the sample solution to the absorbing solution at room temperatures. The temperatures used in conventional evaporation, distillation, sublimation, and oxidation of organic substances are ordinarily 100 to several hundred degrees Celsius. Selective volatilization of impurities from solid or molten samples requires operational temperatures higher than $1000\,^\circ C$.

Contamination from container materials becomes serious at relatively high temperatures used in this method. On the other hand, when the desired trace elements are volatilized, their loss may result from:

(1) imperfect conversion into volatile compounds
(2) slow diffusion in solid samples

Table 12. Volatilization of elements

Volatilized as	Elements volatilized
Simple substance	H, Hg, N, Halogens
Hydride	As, Bi, Cl, F, Ge, N, O, Pb, S, Sb, Se, Sn, Te
Fluoride and oxyfluoride	B, Mo, Nb, Si, Ta, Ti, V, W
Chloride and oxychloride	Al, As, Cd, Cr, Ga, Ge, Hg, Mo, Sb, Sn, Ta, Ti, V, W, Zn, Zr
Bromide	As, Bi, Hg, Sb, Se, Sn
Iodide	As, Sb, Sn, Te
Oxide	As, C, H, Os, Re, Ru, S, Se, Te
Methyl borate	B

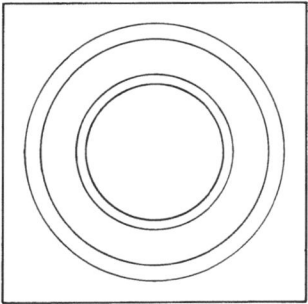

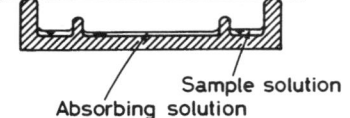

Sample solution
Absorbing solution   Fig. 7. Microdiffusion unit (After Conway [48])

(3) adsorption loss on the apparatus walls, and

(4) imperfect condensation of volatilized compounds.

When the matrices are volatilized, loss of the desired trace elements may result from:

(1) partial volatilization

(2) mechanical entrainment of tiny liquid or solid particles containing the desired elements in the escaping vapor

(3) adsorption on the container walls, and

(4) formation of an acid-insoluble residue containing the desired trace elements.

4.1 Volatilization from Solutions

4.1.1 Volatilization of Trace Elements from Solutions

The desired trace elements in the sample solution are quantitatively volatilized, leaving the matrix elements in the solution, generally with the aid of gas bubbling, heating and chemical reactions. Sample decomposition and the enrichment process are frequently unified. The volatilized compounds are then absorbed in a suitable solution or condensed on a cold solid surface for subsequent determinations. Various types of distillation apparatus are used. An example is shown in Fig. 8. Table 13 lists some applications of this technique, including the well-known Kjeldahl method for the determination of nitrogen in ferrous and nonferrous metals and alloys.

4.1.2 Volatilization of the Matrix from Solutions

The sample solution is simply heated in a dish, beaker or crucible to volatilize the solvent and the matrix elements, leaving the desired trace elements as residues. Small

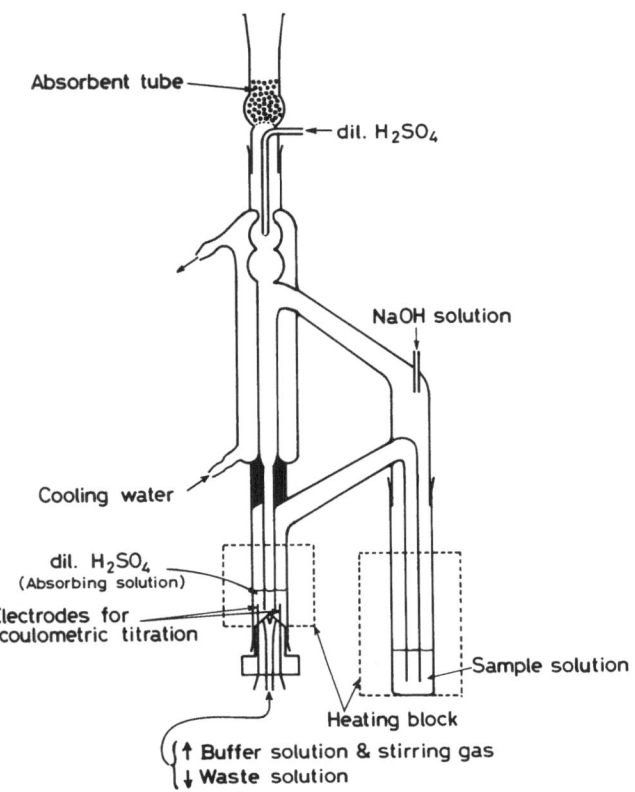

Fig. 8. Closed-circuit distillation apparatus for ammonia (After Werner and Tölg [49])

amounts of reagents such as sulfuric acid, salts and spectrographic grade carbon powder are frequently added to the solution before the volatilization step to minimize the loss of the desired trace elements due to volatilization and adsorption on the container wall and to facilitate redissolution or collection of the residue.

The simplest application of this technique is removal of water, organic solvents, and volatile acids from aqueous or nonaqueous solutions, which is widely used in analyses of water and mineral acids for trace impurities [77–81] and after ion exchange, liquid-liquid extraction and other separations. Loss of trace elements during evaporation to dryness in borosilicate glass dishes was investigated by using radioactive tracers [82]. Boron and heavy metal impurities in silicon tetrachloride are enriched by evaporation of the matrix and determined by spectrophotometry, optical emission spectrometry and atomic absorption spectrometry [83–86]. The product of partial hydrolysis of the matrix silicon tetrachloride acts as suitable collector. See Chap. 5.1.5 for the removal by evaporation of organic solvents after liquid-liquid extraction.

Table 14 tabulates typical examples of selective volatilization of matrix elements, which are generally converted into volatile compounds during the dissolution of the solid sample in acids or organic solvents.

Table 13. Volatilization of trace elements from solutions

Matrices	Trace elements	Volatilized		Determination techniques	Literature
		as	from		
Natural waters	Hg	Hg	Solution + reducing agents (SnCl$_2$, NaBH$_4$)	AAS, OES	[50, 51]
Uranium fluorides and oxides	Br	Br$_2$	H$_2$Cr$_2$O$_7$-H$_2$SO$_4$	Phot.	[52]
Ferrous and non-ferrous metals and alloys	N	NH$_3$	Alkaline solution	Phot., Titr.	[49, 53, 54]
Ni, Ti	S	H$_2$S	HCl, HCl-H$_3$PO$_2$	Phot, Titr., CSV, SIMS	[55–58]
Rocks	Cl	HCl	Strong phosphoric acid	Titr.	[59]
Al, Sea and waste waters	As, Sb	AsH$_3$, SbH$_3$	HCl + Zn	OES	[60, 61]
Natural waters, Rocks, Pb, Ni-base alloys	As, Bi, Ge, Sb, Se, Te	Hydrides	Acidic solution + NaBH$_4$	ICP-OES, AAS	[62–68]
Al, Nb, Zr, Zircaloy	B	BF$_3$	HF-containing acid mixtures	OES	[69]
Pu, U	Si	SiF$_4$	HClO$_4$-HF-HNO$_3$	Phot.	[70]
Te	Sb	SbBr$_3$	H$_2$SO$_4$-HBr-HCl	Phot.	[71]
Te	As, Sn	AsBr$_3$, SnBr$_4$	HClO$_4$-HBr-HCl	Phot.	[72]
Marine sediments	Se	SeBr$_4$	Strong phosphoric acid + NH$_4$Br + KIO$_3$	Phot.	[73]
Be, Ni, Th, U, Zr, Zircaloy	B	B(OCH$_3$)$_3$	Methanol	Phot.	[74–76]

Table 14. Volatilization of matrix elements from solutions

Matrices	Volatilized as	from	Trace elements	Determination techniques	Lit.
Si, SiO$_2$	SiF$_4$	HF-HNO$_3$, HF-HNO$_3$-HClO$_4$, HF-H$_2$SO$_4$	Al, Bi, Cd, Cu, Fe, In, Ni, P, Pb, Tl, Zn	Polar., Phot., Fluor.	[87–89]
B	BF$_3$	HF-HNO$_3$	Al, Ba, Be, Ca, Co, Cr, Cu, Fe, Mg, Mn, Ni, Ti, V, W	OES	[90]
Ge, GeO$_2$	GeCl$_4$	HCl-HNO$_3$-HClO$_4$	Ag, Al, Be, Bi, Cd, Cu, Fe, Ga, In, Mg, Mn, Ni, P, Pb, Sb	OES, Phot.	[91, 92]
Cr	CrO$_2$Cl$_2$	HCl-HClO$_4$	Al, Cu, Fe, Mg, Mn, Ni, Ti, V	OES	[93]
Sn	SnCl$_4$	CCl$_4$-Cl$_2$	Ag, Au, Bi, Co, Cu, Ga, Ni, Pb	OES	[90]
As, GaAs*	AsBr$_3$, AsCl$_3$	HCl-Br$_2$-CCl$_4$	Al, Be, Bi, Cd, Co, Cr, Mg, Mn, Ni, Si, V, Zr	OES	[94, 95]
Sn	SnBr$_4$, SnCl$_4$	HCl-HBr-Br$_2$, HBr-Br$_2$	Cu, Fe, Pb	Phot., Polar.	[96, 97]
Bi	BiBr$_3$	HBr	Pb	Phot.	[98]
Se, SeO$_2$	SeBr$_4$	HBr	Cd, Cu, Fe, Ga, Pb, S, Te, Tl, Alkali and alkaline earth elements	Polar., Turbid., OES	[99–101]
Se, SeO$_2$	SeO$_2$	HNO$_3$-H$_2$SO$_4$, HNO$_3$, H$_2$SO$_4$	Al, As, Ba, Bi, Ca, Cd, Co, Cu, Ga, In, Ni, Pb, Te, Tl	Polar., Phot., OES	[96, 102–107]
B	B(OCH$_3$)$_3$	Methanol	Al, As, Cu, Fe, Mg, Mo, Na, P, Pb, Si	OES	[90]
Al	Aluminum ethyl bromide	C$_2$H$_5$Br	Ag, Co, Cr, Cu, Fe, Mn, Ni, Pb	OES	[108]

* Ga is removed by liquid-liquid extraction.

In *wet oxidation of organic and biological samples* [4], decomposition and removal of the matrix by volatilization are unified. The samples are digested with oxidizing liquid (usually, some combination of sulfuric acid, nitric acid, perchloric acid or hydrogen peroxide) in an open flask, a flask fitted with a reflux condenser, or a closed Teflon or glassy-carbon vessel in a pressure bomb [17]. This technique is often preferred to dry oxidation (see Chap. 4.2.2), because lower temperatures used (usually below 200 °C) and the presence of a large excess of acids result in less evaporation or adsorption loss of the desired trace elements. Under some conditions, however, there still exists the danger of evaporation loss of such elements as As, B, Cr, Ge, Hg, Os, Re, Ru, Sb, Se and Sn. Formation of calcium sulfate precipitates may cause loss of the desired trace elements such as Pb by coprecipitation. Contamination due to reagents, limited sample sizes and careful attention required for the manipulation are disadvantages compared with dry oxidation. Gorsuch recommends wet oxidation in the apparatus shown in Fig. 9 using (1) nitric and sulfuric acids (2) perchloric, nitric and sulfuric acids, and (3) sulfuric acid and hydrogen peroxide [4]. The two-way stopcock permits refluxing, distillation, and removal of the distillate. The desired trace elements finally remain as residue in the flask. A two-necked flask and a thermometer are used when temperature control is required.

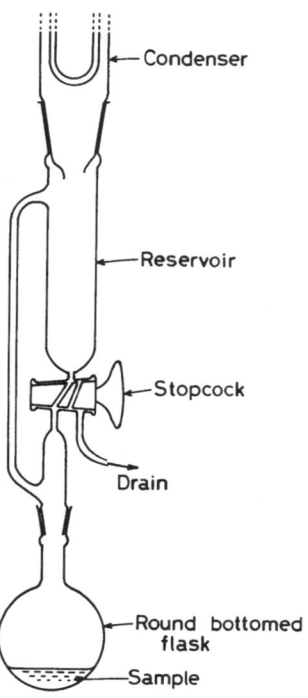

Fig. 9. Apparatus for wet oxidation of organic samples (After Gorsuch [4])

4.2 Volatilization from Solid and Molten States

4.2.1 Volatilization of Trace Elements from Solid and Molten States

Various trace elements can be *selectively volatilized* from solid or molten samples at about 1000 °C or higher temperatures in vacuum or in an inert or reactive gas atmosphere. The volatilized compounds are collected on absorbents, cold traps or condensers (Fig. 10).

This technique is widely used in the determination of H, C, N, O and S in metals and alloys [53, 112]. These elements are volatilized as gaseous compounds from the sample and determined by various gas analytical methods. Hydrogen is quantitatively extracted as molecular hydrogen from a solid metal when the sample is simply heated in vacuum up to the temperature where the diffusion velocity of hydrogen atoms in the metal is appreciably fast. This method is called *hot extraction* or vacuum extraction. Nitrogen and carbon monoxide at the ng/g levels in refractory metals are extracted by levitation melting in ultrahigh vacuum [113]. Hydrogen, N and O are simultaneously extracted into vacuum as molecular hydrogen, molecular nitrogen, and carbon monoxide, respectively, when the sample metal is melted in a graphite crucible by means of high frequency heating. This method is called *vacuum fusion*. The *inert-gas fusion method* uses a stream of inert gas such as argon instead of vacuum. Electrical discharges such as a DC carbon arc in an inert gas atmosphere can be used for the gas extraction instead of high frequency heating. There are several methods using reactive gases. Oxygen is extracted as water vapor by heating the metal sample in hydrogen, in hydrogen sulfide or in hydrogen fluoride, and as sulfur dioxide by heating with sulfur vapor. In the combustion method, where the metal sample is heated in an oxygen or air stream, H, C and S are volatilized as water vapor, carbon dioxide and sulfur oxides, respectively [114–116]. Carbon is also extracted as carbon disulfide from metallic silicon and germanium by treating with sulfur vapor at about 1000 °C [117].

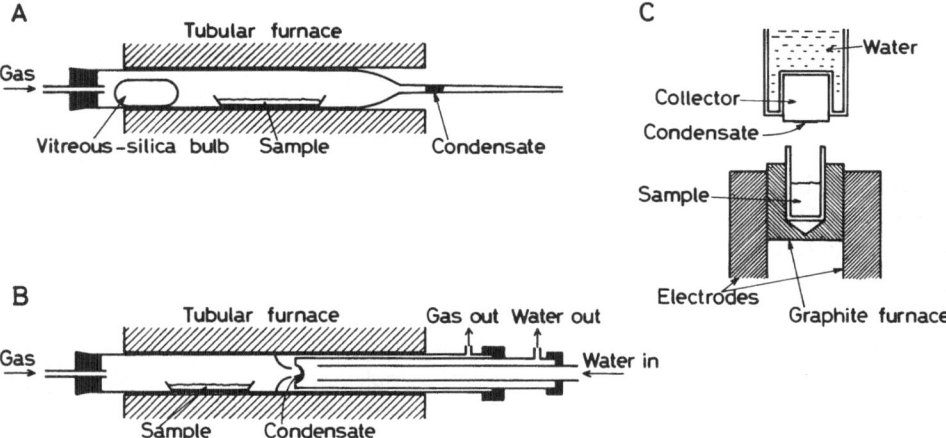

Fig. 10. Air- and water-cooled condensers. A: After Geilmann and Neeb [109], B: After Heinrichs [110], C: After Zil'bershtein [111]

Traces of Se in metals, alloys, rocks and other matrices are evaporated as selenium dioxide in a stream of oxygen or argon-oxygen mixtures, and collected in a liquid nitrogen trap for atomic absorption spectrometry [118, 119]. Similarly Bi, Cd, Pb and Tl are quantitatively volatilized [120]. In pyrohydrolysis, traces of B, F and Cl in metals, rocks and other matrices are volatilized as boron trifluoride, hydrogen halides, etc. by heating the sample in a stream of water vapor or damp gas [3, 121]. Such impurities as Al, Be, Co, Fe, Ga, In, Mn, Ni, Sn and Ti in synthetic silicon dioxide or natural quartz are volatilized by heating in a stream of hydrogen chloride and condensed on a water-cooled carbon collector for optical emission spectrography [111]. Many similar techniques are described for the enrichment of Zn and other trace elements in metals, rocks and other matrices [109, 110, 122–129].

Volatilization of trace metals is also carried out *in vacuum* or in air instead of in a stream of reactive or inert gases. Traces of Al, Ca, Cu, Mg, Ti and Zn in high-purity silicon are evaporated in vacuum [130]. Various metallic and nonmetallic impurities in refractory powder materials such as oxides of uranium, aluminum, thorium, zirconium and beryllium are evaporated at 1500 to 2000 °C in an electrically heated crucible in vacuum or in air, and condensed on a water-cooled metallic or graphite rod (Fig. 10 C), which is then used as an electrode for emission spectrography [111, 131, 132]. A similar selective volatilization process takes place during the DC carbon arc excitation of refractory matrices such as uranium oxide in the presence of a carrier (e.g. Ga_2O_3) for emission spectrography. This method, called *carrier distillation*, is regarded as a unification of enrichment and determination steps.

4.2.2 Volatilization of the Matrix from Solid and Molten States

Removal of water by sublimation in vacuum from frozen samples is called *lyophilization* or *freeze-drying*, which is useful for drying biological tissues as well as in water analysis [133, 134]. Such impurities as Cu, Fe, Ni and Pb in commercial grade ammonium chloride are enriched by *sublimation* of the matrix for the emission spectrographic determination [135]. A small amount of 1% phosphoric acid is added before sublimation to retain the desired trace elements in the residue. Alkali carbonate and nitrate matrices are removed by volatilization in a stream of argon, and traces of Co, Cr, Cu, Fe, Mn and Ni in the residue are determined by flameless atomic absorption spectrometry [136, 137]. Traces of C in sodium is enriched by removal of the matrix by vacuum distillation and determined by the combustion method [138]. Such impurities as Ag, Bi, Cu and Pb in high-purity cadmium are determined by emission spectrography or polarography after steam distillation of cadmium [139].

Graphite is ashed in the presence of calcium hydroxide in a stream of oxygen and B in the ash is determined by spectrophotometry after distillation from methanol [140]. Removal of the matrix by combustion is also useful in the enrichment of impurities in high-purity sulfur and selenium [141, 142].

Dry oxidation of organic and biological samples [4] is usually carried out by heating the sample at 450 to 500 °C in air. Although this technique is widely used because of its simplicity, the following elements may be lost by volatilization as metals, chlorides or organometallic compounds under certain conditions, according to their chemical forms in the samples as well as chemical reactions during the oxidation: As,

B, Cd, Cr, Cu, Fe, Hg, Ni, P, Pb, V and Zn. Also, nonmetals of periodic groups IV, V, VI, VII (except S and P) and possibly Ga, In and Tl cannot be recovered in dry oxidation [22]. Losses also may occur due to adsorption on the vessel and formation of acid-insoluble residues. There are a number of modified dry oxidation procedures. Addition of solid salts or moistening the sample with salt solutions may be useful in some cases in preventing losses of the desired trace elements. Other useful procedures include combustion in an oxygen bomb or an oxygen flask and so-called low-temperature ashing in high-frequency-excited oxygen plasmas. A standard procedure for dry oxidation [4, 22] is given below, although minor modifications are sometimes required.

A 5- to 10-g sample is placed in a suitable vitreous-silica or platinum dish, thinly spread over a fairly large area. If appropriate, an ashing aid (e.g. 10 ml of 10% sulfuric acid or 7% magnesium nitrate solution) is added to facilitate the decomposition and/ or to improve the trace recovery. The sample is dried and thoroughly charred in an evaporation chamber (Fig. 5), and then heated in a furnace at about 450 °C (sometimes at about 500 °C when magnesium nitrate is used as ashing aid) overnight, or for a similar period. The ignition of the sample must always be avoided. If unoxidized organic matter remains, the residue is moistened with water or nitric acid (1 + 2), and after evaporation to dryness, again heated in the furnace for a further period. The resulting ash is moistened with a little water followed by 10 ml of hydrochloric acid (1 + 1) and evaporated to dryness. The residue is then dissolved in hydrochloric acid (1 + 9) or other suitable solvent.

Various matrices are *volatilized as halides or oxyhalides* by reactions with halogen-containing gases: e.g. Si, SiO_2 and TiO_2 with hydrogen fluoride; As, Ga, Sb, Sn, Ti and Zr with chlorine; Al, Ta, V_2O_5, MoO_3 and WO_3 with hydrogen chloride; and TiO_2 with carbon tetrachloride [111]. The chlorination technique has been applied to enrichment of traces of Li, Na, K and Ca in phosphorus, arsenic and antimony [143], Ca in zirconium and its alloys [144] and Ca in aluminum and tantalum [145] for flame photometry. For the isolation of stable oxides such as silicon and aluminum oxides in steels [53, 146], the sample as such or the residue from the anodic dissolution of the sample is chlorinated.

A review is available on the separation of trace elements in solid samples by formation of volatile inorganic compounds [147].

5 Liquid-Liquid Extraction

The enrichment techniques discussed in this chapter are based on the distribution of solutes between two essentially immiscible solvents (see Appendix A. 1). Extraction of various metal elements from an aqueous solution into an organic solvent such as ether, chloroform and carbon tetrachloride, is most widely used in inorganic trace analysis. The formation of an uncharged chemical species by chelation and ion association is essential for the extraction. Covalent molecules are also extracted into organic solvents; e.g. I_2, halides of As(III), Ge(IV), etc., oxides of Os(VIII) and Ru(VIII), and piazselenol. Several monographs are available on liquid-liquid extraction [14, 148–153].

In the case of the distribution of a single chemical species as solute between the two liquid phases, the partition constant, P, which is defined as the ratio of concentration of the solute in the organic phase to that in the aqueous phase at equilibrium, is independent of the concentrations, provided that the activity coefficients of the solute in the two phases remain constant. Generally, several species of an element exist as solutes in both phases. Therefore, the distribution ratio, D, of an element is defined as the ratio of the total concentration of the element in the organic phase to that in the aqueous phase at equilibrium. The distribution ratio is often dependent on the total concentration of the element. When the desired trace element is extracted, the larger the distribution ratio of the trace element and the smaller that of the matrix, the higher the trace recovery and the enrichment factor are obtained. When the matrix is removed by the extraction, the reverse is required for the successful enrichment. The distribution ratio of an element depends greatly on chemical reactions including cheletion, ion association, dissociation, polymerization and solvation, in the two phases. Therefore, proper choice of extraction reagents and their concentration, organic solvents, pH or acidity of the aqueous phase, masking agents and salting-out agents is important.

5.1 General Procedures

5.1.1 Batch Extraction

Usually, an aqueous sample solution and an organic solvent are thoroughly contacted in a pear-shaped separatory funnel by shaking it by hand or with a mechanical shaker. After establishment of equilibrium, the two phases are allowed to separate, and then

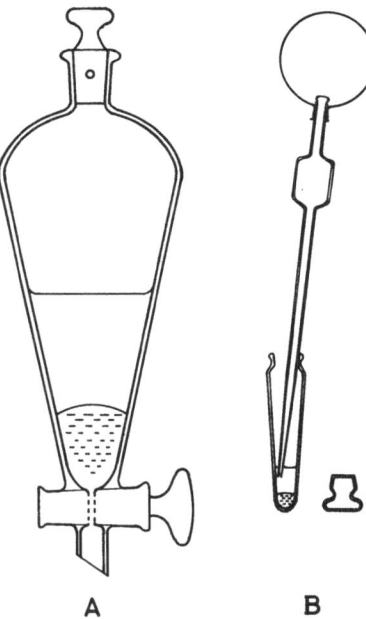

Fig. 11. Apparatus used in batch extraction.
A: Separatory funnel, B: Centrifuge tube

the lower phase is drained off through a stopcock (Fig. 11 A). Extractions with small volumes of a sample solution and an organic solvent are carried out in a small container such as a glass-stoppered centrifuge tube (Fig. 11 B).

When a trace element having a distribution ratio D is extracted from V ml of an aqueous phase into V_o ml of an organic phase, the percent extraction E (%), i.e. the trace recovery, is

$$E\,(\%) = 100 \times \left(1 + \frac{1}{D}\frac{V}{V_o}\right)^{-1} = 50\left[1 + \tanh\left\{\frac{2.303}{2}\left(\log D + \log \frac{V_o}{V}\right)\right\}\right] \qquad (3)$$

which is depicted in Fig. 12. If the distribution ratio is not large enough, batch extractions are repeated twice or more with portions of fresh solvent, and the organic phases are then combined. The percentage of the trace element remaining in the aqueous phase after n extractions is $100[1 + D(V_o/V)]^{-n}$ (%). For a given total amount of solvent (nV_o = constant), the best results are obtained by a relatively large number of extraction with small amounts of solvent. In reality, the situation is somewhat more complicated. Mutual dissolution of the two phases occurs more or less during the extraction, and the volumes of the two phases and the distribution ratio of the trace element are changed during the repeated extractions. Pre-equilibration, i.e. preliminary shaking of each phase with the other solvent, can minimize these phenomena.

Sometimes, vigorous shaking is harmful because of the formation of emulsion which makes the separation of the two phases difficult. Small amounts of water droplets accompanying the organic solvent after the phase separation are removed by

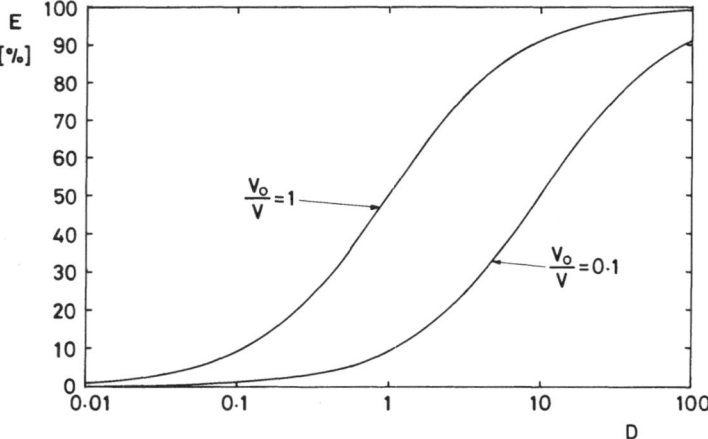

Fig. 12. Percent extraction vs. distribution ratio

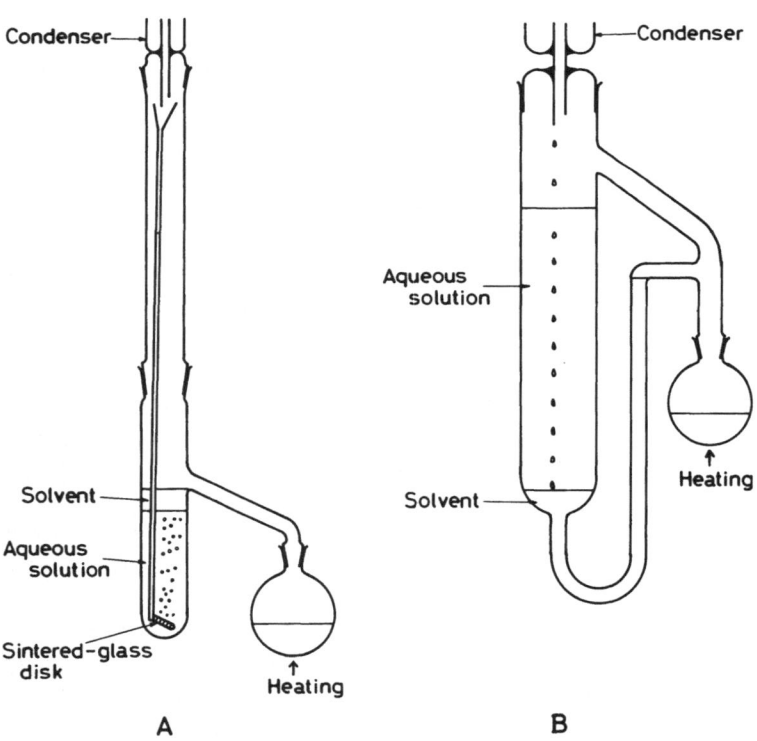

A   B

Fig. 13. Continuous extractors. A: For solvents lighter than water, B: For solvents heavier than water

filtration through a dry filter paper or a Teflon membrane filter, by adding anhydrous sodium sulfate, or by centrifugation.

It must be kept in mind that trace elements may sometimes be lost by adsorption on the wall of the separatory funnel, solid particulates or the liquid-liquid interface.

5.1.2 Continuous Extraction

This technique is useful, when the distribution ratio is so small that repeated batch extractions are not applicable. There are many types of continuous extractors: in some of them the spent solvent is recycled as shown in Fig. 13.

When dV_o ml of an organic solvent is passed through V ml of an aqueous solution containing Q g of an element,

$$-dQ = kQ \, dV_o/V \tag{4}$$

where k is a constant which closely approaches the distribution ratio if the value of the latter is low. Integrating Eq. (4), we obtain

$$Q = Q_0 \exp(-kV_o/V) \tag{5}$$

where Q_0 g is the initial amount of the element. The half-extraction volume, the volume of the solvent required to reduce the amount of the element to one-half, is therefore 0.693 V/k.

5.1.3 Countercurrent and Chromatographic Extractions

These techniques are applicable to the cases where the distribution ratios of the elements to be separated each other are of the same order of magnitude.

The principle of discontinuous countercurrent extraction is as follows. Two series of tubes, as shown in Fig. 14, are used. The upper and lower tubes contain equal volumes of an organic solvent and an aqueous solution, respectively. The contents of tube L_0 containing the solute and tube U_0 are equilibrated, and after the separation of the two phases, the upper tubes are shifted so that U_0 is over L_1 and U_1 is over L_0. Then the contents of tubes U_0 and L_1 and of tubes U_1 and L_0 are equilibrated, respectively, and after the separation of the two phases, the upper tubes are shifted so that U_0 is over L_2, U_1 is over L_1, and U_2 is over L_0. After n transfers, the fraction $T_{n,r}$ of the solute present in the $L_r - U_{n-r}$ pair is given by Eq. (6).

$$T_{n,r} = \frac{n!}{r! \, (n-r)!} \left(\frac{D}{1+D}\right)^r \left(\frac{1}{1+D}\right)^{n-r} \tag{6}$$

This is the binomial distribution with an average of $nD/(1 + D)$ and a standard deviation of $\sqrt{nD/(1 + D)^2}$, which is approximated by Gaussian normal distribution with the same average and standard deviation when n is sufficiently large. Distribution curves are shown in Fig. 14. Satisfactory separations can be achieved at large n. Automatic apparatus used for this separation technique are available.

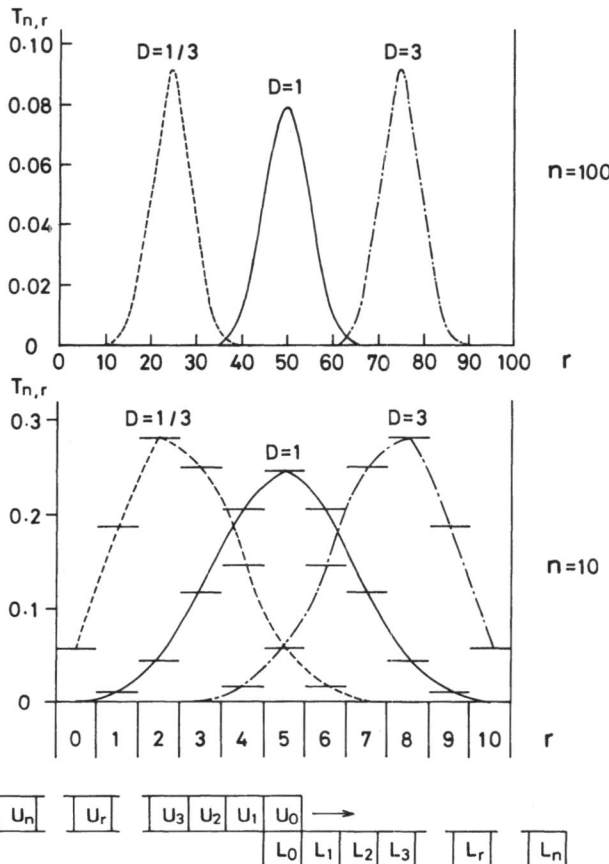

Fig. 14. Discontinuous countercurrent extraction

Similarly, successful separations can be achieved by *partition liquid chromatography*, which uses partitioning of the solute between a stationary liquid phase supported on a suitable inert solid and a moving liquid phase (see Chap. 9). In *droplet countercurrent chromatography*, a large number of droplets of moving phase are passed through a series (say, 300) of small-bore columns containing a stationary phase (Fig. 15); turbulence within the droplet promotes efficient partitioning of the solute between the two phases [154].

5.1.4 Backwashing

After the extraction, the organic phase may contain small amounts of matrix elements, extracted together with the desired trace elements or contained in small water droplets. To remove these, the organic phase is shaken with one or more small portions of an aqueous solution containing suitable reagents, so that the matrix elements are selectively back-extracted or removed into the aqueous phase. Under proper conditions, practically no loss of the desired trace elements occurs. This technique, which is called backwashing, is useful to improve enrichment factors.

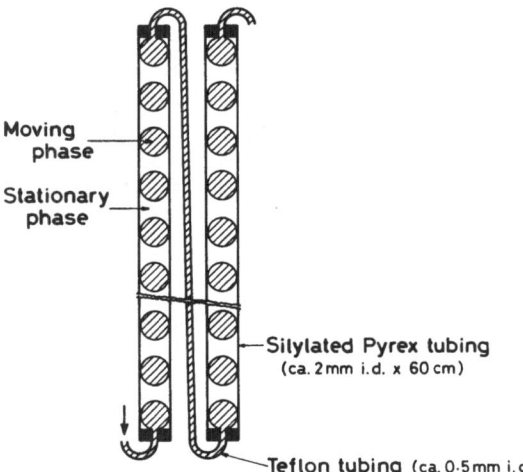

Fig. 15. Droplet countercurrent chromatography (After Tanimura et al. [154])

5.1.5 Stripping

Organic solvents containing the desired trace elements can be directly delivered to a number of determination techniques including spectrophotometry, optical emission and atomic adsorption spectrometry, and radioactivity measurements. Enhancement of signals in the presence of organic solvents is well known in flame emission and absorption spectrometry. However, it is frequently necessary to remove the extracted desired trace elements from the organic phase to an aqueous phase before determination steps. This process, called stripping, is carried out by the following two techniques.

One is the *back-extraction* of the desired trace elements into an aqueous solution containing acids or other reagents under conditions whereby extractable complexes are destroyed. The separation of the extracted elements from each other is often effected in the back-extraction.

The other technique is the *evaporation of the organic solvent,* usually in the presence of small amounts of water and mineral acids, followed by dry or wet oxidation of residual organic complexes and dissolution of the residue in an aqueous solution. This is very simple, but care must be taken to avoid loss of the desired trace elements due to volatilization and danger due to the inflammability of organic solvents.

5.2 Extraction of Metal Chelates

5.2.1 Chelate Extraction Systems

Various chelating agents are useful for extracting metal elements as uncharged metal chelates into organic solvents from aqueous solutions. Typical chelate extraction systems are listed in Table 15.

Table 15. Typical chelate extraction systems [152]

Chelating agents	Metal chelates	Elements extracted
Oxine (8-Hydroxyquinoline, 8-Quinolinol)		Ag, Al, Am, Ba, Be, Bi, Bk, Ca, Cd, Ce, Cf, Cm, Co, Cr, Cu, Er, Fe, Ga, Hf, Hg, In, La, Mg, Mn, Mo, Nb, Nd, Ni, Np, Pa, Pb, Pd, Pu, Sb, Sc, Sm, Sn, Sr, Th, Ti, Tl, U, V, W, Y, Zn, Zr
Dithizone (Diphenylthiocarbazone)		Ag, Au, Bi, Cd, Co, Cu, Fe, Ga, Hg, In, Mn, Ni, Pb, Pd, Po, Pt, Sn, Te, Tl, Zn
Sodium diethyldithiocarbamate (DDTC)		Ag, As, Au, Bi, Cd, Co, Cr, Cu, Fe, Ga, Hg, In, Mn, Mo, Ni, Pb, Pd, Pu, Sb, Se, Sn, Te, Ti, Tl, U, V, W, Zn (Ir, Nb, Os, Po, Pt, Rh, Ru)*
Ammonium pyrrolidine-dithiocarbamate (APDC, Ammonium tetramethylene-dithiocarbamate)		Ag, As, Au, Bi, Cd, Co, Cr, Cu, Fe, Ga, Ge, Hg, In, Ir, Mn, Mo, Nb, Ni, Os, Pb, Pd, Pt, Re, Rh, Ru, Sb, Se, Sn, Tc, Te, Tl, U, V, W, Zn
Cupferron (Ammonium salt of nitrosophenyl-hydroxylamine)		Al, Bi, Ce, Co, Cu, Fe, Ga, In, Mo, Nb, Pa, Pb, Pd, Sb, Sn, Th, Ti, U, V (Ag, Cd, Hf, Hg, La, Mn, Ni, Sc, Tl, W, Y, Zn, Zr)*
Acetylacetone		Al, Be, Bi, Co, Cr, Cu, Fe, Ga, Hf, Hg, In, Mn, Mo, Pb, Pd, Pu, Ru, Sc, Sn, Th, Ti, Tl, U, V, Zn, Zr
Thenoyltrifluoroacetone (TTA)		Ac, Al, Am, Be, Bi, Bk, Ca, Ce, Cf, Cm, Co, Cr, Cs, Cu, Es, Eu, Fe, Fm, Hf, In, La, Mn, Mo, Ni, Np, Pa, Pb, Pd, Po, Pt, Pu, Sc, Sn, Sr, Th, Tl, U, W, Y, Yb, Zr
1-(2-Pyridylazo)-2-naphthol (PAN)		Ag, Bi, Cd, Ce, Co, Cu, Eu, Fe, Ga, Hg, In, Ir, La, Mn, Ni, Pb, Pd, Pt, Rh, Sc, Sn, Th, Ti, U, V, Y, Zn, Zr

* Partial extraction

5.2.2 Equilibria in Chelate Extraction Systems

Distribution curves for the chelating agent. Oxine, an amphoteric chelating agent (HR), is taken as an example. There exist the following equilibria in the aqueous phase.

$$R^- + H^+ \rightleftharpoons HR, \qquad K = \frac{[HR]}{[R^-][H^+]} = 7 \times 10^9 \tag{7}$$

$$HR + H^+ \rightleftharpoons H_2R^+, \quad K' = \frac{[H_2R^+]}{[HR][H^+]} = 1.3 \times 10^5 \tag{8}$$

The species HR is extracted into the organic phase. Thus,

$$HR \rightleftharpoons (HR)_o, \qquad P_r = \frac{[HR]_o}{[HR]} = 720 \tag{9}$$

where subscript o refers to organic phase. The distribution ratio of the reagent then becomes

$$D = \frac{[HR]_o}{[H_2R^+] + [HR] + [R^-]} = \frac{P_r}{K'[H^+] + 1 + (K[H^+])^{-1}} \tag{10}$$

Therefore,

$$D = P_r \ . \tag{11}$$

in neutral solutions,

$$\log D = pH + \log P_r - \log K' \tag{12}$$

in acidic solutions, and

$$\log D = -pH + \log P_r + \log K \tag{13}$$

in basic solutions (Fig. 16). The sigmoidal parts of the E-pH curves in Fig. 16 are approximated by Eqs. (14) and (15), Eqs. (3), (12) and (13) being used.

$$E = 50 \left[1 + \tanh \left\{ \frac{2.303}{2} \left(pH + \log \frac{P_r V_o}{K' V} \right) \right\} \right] \tag{14}$$

$$E = 50 \left[1 - \tanh \left\{ \frac{2.303}{2} \left(pH - \log \frac{P_r K V_o}{V} \right) \right\} \right] \tag{15}$$

The pH value at D = 1 (i.e. E = 50% at $V_o/V = 1$) is generally designated as $pH_{1/2}$.

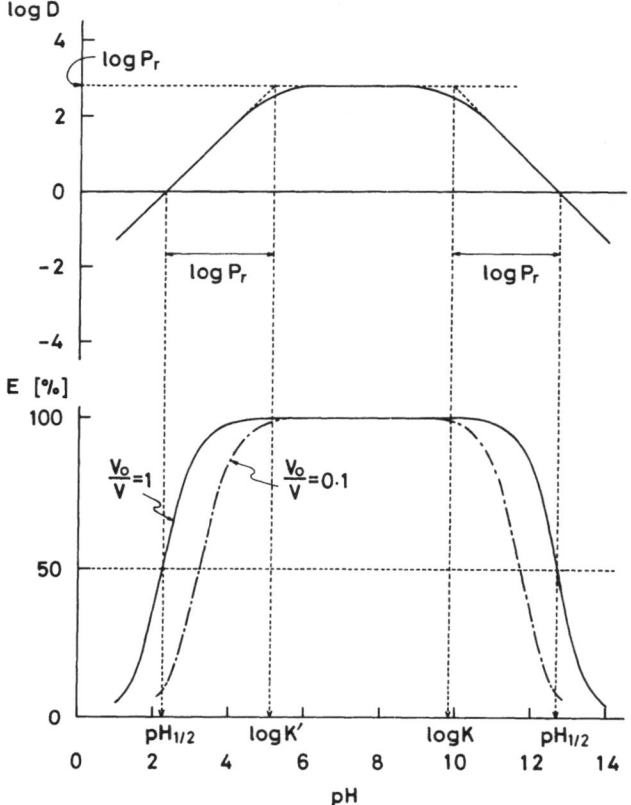

Fig. 16. Distribution of oxine between water and chloroform

Distribution curves for the metal element. The extraction of the metal element from an aqueous solution to an organic solvent is expressed as

$$M^{n+} + n(HR)_o \rightleftharpoons (MR_n)_o + nH^+, \quad K_{ex} = \frac{[MR_n]_o\,[H^+]^n}{[M^{n+}]\,[HR]_o^n} \tag{16}$$

where K_{ex} is the extraction constant. For the formation of the chelate in the aqueous phase

$$M^{n+} + nR^- \rightleftharpoons MR_n, \qquad \beta = \frac{[MR_n]}{[M^{n+}]\,[R^-]^n} \tag{17}$$

and for the distribution of the chelate between the two phases

$$MR_n \rightleftharpoons (MR_n)_o, \qquad P_c = \frac{[MR_n]_o}{[MR_n]} \tag{18}$$

Using Eqs. (7), (9), (16), (17) and (18), we may write for K_{ex}

$$K_{ex} = \beta \, P_c P_r^{-n} K^{-n} \tag{19}$$

which means the constant K_{ex} is dependent on the metal element, chelating agent and organic solvent.

Assuming that M^{n+} and MR_n are the only chemical species containing the metal element in the aqueous and organic phases, respectively, we obtain, using Eq. (16),

$$D = \frac{[MR_n]_o}{[M^{n+}]} = K_{ex} \frac{[HR_n]_o^n}{[H^+]^n} \tag{20}$$

or

$$\log D = \log K_{ex} + n \log [HR]_o + n \, pH \tag{21}$$

As D becomes very large, $[MR_n]$ greatly exceeds $[M^{n+}]$ in the aqueous phase, and D approaches to the limiting value D_{lim}.

$$D_{lim} = \frac{[MR_n]_o}{[MR_n]} = P_c \tag{22}$$

The distribution ratio is independent of the metal concentration. The pH dependence of log D or E is shown in Fig. 17, when $[HR]_o$ is constant. From Eqs. (3) and (21), an approximate equation for the E-pH curve may be derived.

$$E = 50 \left[1 + \tanh \left\{ \frac{2.303}{2} \left(n \, pH + n \log [HR]_o + \log \frac{K_{ex} V_o}{V} \right) \right\} \right] \tag{23}$$

From Eq. (21), we obtain

$$pH_{1/2} = -(\log K_{ex})/n - \log [HR]_o \tag{24}$$

According to Eq. (21), the gradients of the log D-log $[HR]_o$ curve at constant pH and the log D-pH curve at constant $[HR]_o$ are both n. When the extracted chemical species is $MR_n \cdot (m - n) HR$, we have, instead of Eq. (21),

$$\log D = \log K_{ex} + m \log [HR]_o + n \, pH \tag{25}$$

which gives different gradients, m and n, for the two curves. For example, when VO_2^+-oxine complexes are extracted from acidic solutions, m = 2 and n = 1 at a small excess of oxine, and m = 3 and n = 1 at a large excess of oxine.

In the above discussion on the pH dependence of log D or E, it is assumed that $[HR]_o$ is constant. When a given amount of the chelating agent is used, $[HR]_o$ varies with pH as shown in Fig. 16. At pH $\ll$ $pH_{1/2}$ = log K' $-$ log P_r, we obtain, from Eq. (10),

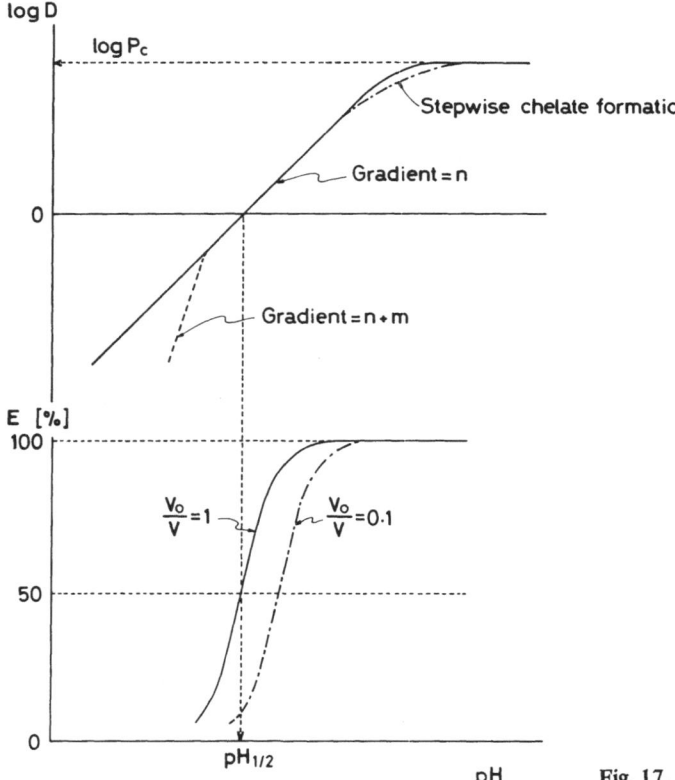

Fig. 17. Extraction of metal chelate

$$[HR]_o = C_r P_r / (K' [H^+])$$ (26)

where C_r is the total concentration of the chelating agent in the aqueous phase. Combining Eqs. (25) and (26), we find

$$\log D = m \log C_r + (n + m) pH + const.$$ (27)

Therefore, the gradient of the log D-pH curve may change between n and (n + m) as shown in Fig. 17.

Sometimes stepwise formation of chelates must be considered. In this case,

$$D = \frac{[MR_n]_o}{[M^{n+}] + [MR^{(n-1)+}] + [MR_2^{(n-2)+}] + \ldots \ldots + [MR_n]} = \frac{[MR_n]_o}{[M^{n+}] \alpha_{M(R)}}$$ (28)

where $\alpha_{M(R)} = 1 + K_1 [R^-] + K_1 K_2 [R^-]^2 + \ldots + K_1 K_2 \ldots K_n [R^-]^n$, $K_1, K_2, \ldots$, K_n are stepwise (or consecutive) formation (or stability) constants, and $K_1 K_2 \ldots K_n = \beta$. Therefore we obtain

$$\log D = \log K_{ex} + m \log [HR]_o + n pH - \log \alpha_{M(R)}$$ (29)

Since HR is a weak acid, $\alpha_{M(R)}$ increases with increasing pH and thus log D changes with pH as shown in Fig. 17.

As the pH of the aqueous phase increases, D and E increase approaching P_c and 100%, respectively. At higher pH, however, D and E decrease again due to hydrolysis of metal ions and formation of water-soluble chelates such as MR_{n+1}^{-1}. Fig. 18 shows extraction curves for several metal oxinates [155].

5.2.3 Masking [156, 157]

When two metal elements form extractable chelates with a particular chelating agent, a large difference in $pH_{1/2}$ values of both elements is required to achieve successful mutual separation at an appropriate pH.

By adding a masking agent L, $pH_{1/2}$ can be shifted. Most commonly used masking agents include cyanide, tartrate, citrate, fluoride and EDTA (see Appendix A. 2). They prevent the extraction of metal elements into the organic phase by forming strong water-soluble (usually negatively charged) complexes. In the presence of L,

$$D = \frac{[MR_n]_o}{[M^{n+}]\,\alpha_{M(L)}} \tag{30}$$

where $\alpha_{M(L)} = 1 + K_1'\,[L^-] + K_1'\,K_2'\,[L^-]^2 + \ldots . + K_1'\,K_2'\ldots . K_n'[L^-]^{n'}$, and K_1', K_2', $\ldots$, $K_{n'}$ are stepwise formation constants. Therefore, we obtain

$$\log D = \log K_{ex} + m \log [HR]_o + n\,pH - \log \alpha_{M(L)} \tag{31}$$

Thus $pH_{1/2}$ increases by $(1/n) \log \alpha_{M(L)}$. Since HL is a weak acid, $\alpha_{M(L)}$ increases with increasing pH. The pH dependence of log D in the presence of a masking agent is generally rather complicated and a maximum and a minimum sometimes appear in a log D-pH curve.

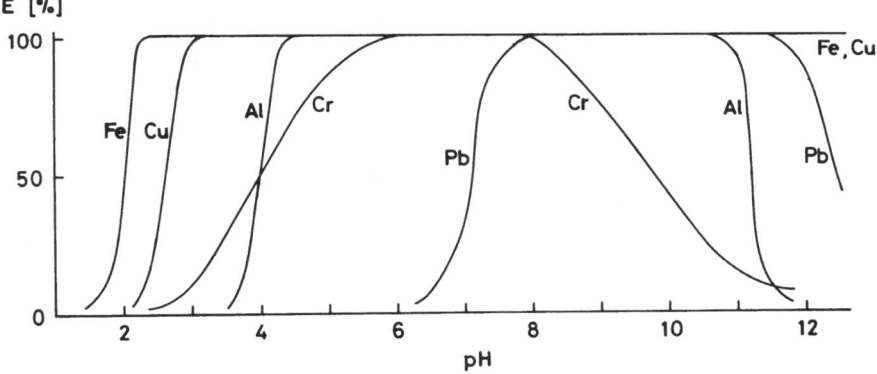

Fig. 18. Extraction of metal oxinates [155] $V_o/V = 0.1$

Table 16. Use of masking agents in the dithizone extraction [152]

Masking agents	Solutions	Metals reacting
Cyanide	Basic solutions	Bi, In, Pb, Po, Sn(II), Tl(I)
Cyanide	Slightly acidic solutions	Ag, Cu, Hg(II), In, Pd(II), Po
Bromide or iodide	Dilute acidic solutions	Au, Cu, Pd
Thiocyanate	Dilute acidic solutions	Au, Cu, Hg
Thiocyanate plus cyanide	Dilute acidic solutions	Cu, Hg
Thiosulfate	Slightly acidic solutions	Cd, Co, In, Ni, Pd, Sn(II), Zn
Thiosulfate plus cyanide	Slightly acidic solutions	Sn(II), Zn
EDTA	Dilute acidic solutions	Ag, Hg
DDTC	Basic solutions	Zn
DDTC	Slightly acidic solutions	Bi, Cd, Zn
Bis(2-hydroxyethyl)-dithiocarbamate	Slightly basic solutions	Zn

Selectivity in chelate extraction systems can be much improved by simultaneously using two or more masking agents. A large number of elements are extracted into chloroform or carbon tetrachloride with dithizone (Table 15), but use of masking agents and adjustment of pH or acidity of the aqueous phase restrict the number of extracted elements as shown in Table 16.

Other means for improving selectivity include use of an appropriate metal chelate instead of a chelating agent itself [158] and the substoichiometric technique described in Chap. 5.2.7.

5.2.4 Synergism

Synergism is generally defined as the joint action of two agents to produce an effect greater than the sum of the independent effects. In metal chelate extraction systems, synergic enhancement of the distribution ratio often appears when two extraction reagents react with the metal to form a more easily extractable species compared with simple chelates. For example, magnesium forms $Mg(Ox)_2(Bu)_2$ with oxine (Ox) and n-butylamine (Bu), and europium(III) forms $Eu(TTA)_3(TBP)_2$ with thenoyltrifluoroacetone (TTA) and tributyl phosphate (TBP).

5.2.5 Coextraction

Coextraction is analogous to coprecipitation in precipitation processes (see Chap. 7.1.1). It implies the simultaneous extraction of an element together with another element under the conditions where the former is not extracted in the absence of the latter. For example, calcium and strontium are coextracted with scandium, neodymium and thorium into a benzene solution of oxine, presumably due to the formation of extractable species such as $Ca(ScOx_4)_2$. Coextraction, however, occurs less frequently than coprecipitation.

5.2.6 Extraction Rate

The rate of extraction of metals as chelates from aqueous solutions into organic solvents is governed by the rate of various chemical reactions in the two phases as well as that of material transport across the interface of the two phases. In most chelate extraction systems, the equilibrium is reached within a few min or less, when the two phases are shaken vigorously. Some extractions, however, require hours or more for attainment of equilibrium. Examples include extractions of

Bi(III), Co(II), Cu(II), Ga(III), Tl(I) and Zn(II) with dithizone,

of Be(II), Fe(III) and Pu(IV) with TTA, and

of Co(II), Cr(III), Fe(III), Mg(II), Mo(VI) and Ni(II) with acetylacetone.

In these cases, the formation of chelates is considered as a rate-determining step. The slow extraction rate is also observed in some back-extractions.

By taking advantage of the great difference in extraction rate, selective extraction is achieved by limiting the shaking to a short period. Thus, Hg(II) and Au(III) are separated from Cu(II) in the dithizone extraction.

5.2.7 Chelate Extraction of Trace Elements

Various trace heavy metals in natural waters, waste waters, and other water samples are simultaneously extracted at appropriate pH as chelates into a small volume of an organic solvent, leaving alkali and alkaline earth elements in the aqueous phase. Some examples are listed in Table 17. The concentrations of the trace elements at the μg/l or ng/l level in the samples are increased by one to two orders of magnitude in the organic phase, which is delivered to atomic absorption spectrometry or other determination techniques, as such or after stripping. The trace concentrations are further increased by back-extraction with a smaller amount of aqueous solution. It must be kept in mind, however, that the trace recoveries are often low in direct extractions from natural waters, because part of the desired trace elements exists in chemical forms which do not react with the chelating agents (see Chap. 12). Although most of the extractions are carried out in a separatory funnel using a 10- to 1000-ml sample, the discontinuous countercurrent extraction (n = 400) using a carbon tetrachloride solution of dithizone or chloroform solution of oxine was applied to concentrate Ag, Al, La, Mn, Mo, Ni, Pb, Sn, V and Zn in 8 liters of sea water for the spectrographic determination [177]. An automated extraction procedure is described for the enrichment of traces of Co, Cr, Cu, Fe, Mn and Zn in sea water [178].

There are many applications of this technique in the enrichment of trace elements at the low μg/g or the ng/g level in high-purity materials and other inorganic solid samples. After dissolution of the samples, the desired trace elements are selectively extracted into the organic phase, leaving the matrix elements in the aqueous phase. Table 18 lists some examples.

Trace metal elements in *biological samples* such as blood, urine and tissues, are directly, or after oxidation of the samples, extracted as chelates from aqueous solutions into organic solvents prior to the atomic absorption spectrometric determination [195—198].

The chelate extraction has wide applicability in rapid *radiochemical separations* with or without the carrier. Ordinarily, an excess of a chelating agent is used to

Table 17. Chelate extraction of trace elements in water samples

Chelating agents	Organic solvents	Trace elements	Determination techniques	Lit.
Oxine	Chloroform	Mn	AAS	[159]
Dithizone	Chloroform, Nitrobenzene	Ag, Cd, Co, Cu, Ni, Pb, Zn	AAS	[160–162]
DDTC	Diisobutyl ketone, 3-Methyl-1-butanol, Chloroform, Acetone-chloroform	Au, Cd, Co, Cu, Fe, Hg, Mn, Pb, U, Zn	AAS, NAA, Fluor., HPLC	[163–167]
APDC	MIBK, Diisobutyl ketone, Carbon tetrachloride	Ag, Cd, Co, Cr, Cu, Fe, Mn, Mo, Ni, Pb, V, Zn	AAS	[168–172]
APDC + diethyl-ammonium diethyldithio-carbamate	1,1,2-Trichloro-1,2,2-trifluoro-ethane, Chloroform	Cd, Co, Cu, Fe, Mo, Ni, Pb, V, Zn	AAS, ICP-OES	[173, 174]
Hexamethylene-ammonium hexamethylene-dithiocarbamate	Diisopropyl ketone + xylene	Ag, Bi, Cd, Co, Cu, Ni, Pb, Tl, Zn	AAS	[175]
PAN	MIBK	Co, Cu, Fe, Ni, Zn	AAS	[176]

extract the desired trace elements as much as possible by a single batch operation. In the isotope dilution method, however, a substoichiometric amount of a chelating agent is successfully used to extract a constant and extremely small amount of a metal element for radioactivity measurement. This simple technique makes the isotope dilution method highly sensitive and selective [199]. It is also useful in radiochemical separations for neutron activation analysis.

5.2.8 Chelate Extraction of Matrix Elements

Extraction of matrix elements as chelates into the organic phase, leaving the desired trace elements in the aqueous phase, has limited applicability because of moderate solubility of chelates in organic solvents with some exceptions. In addition, the matrix extraction generally is not preferable to the trace extraction, because of hazards of loss of the desired trace elements due to coextraction and contamination due to larger amounts of chelating agents used. This technique, however, is successfully applied in the enrichment of traces of Mn in high-purity niobium, tantalum, molybdenum and tungsten metals by extraction of the matrix elements with cupferron into chloroform or chloroform-3-methyl-1-butanol [200].

Table 18. Chelate extraction of trace elements in high-purity materials and other inorganic solid samples

Matrices	Trace elements	Chelating agents	Organic solvents	Determination techniques	Lit.
$NaHCO_3$	Mg	Oxine	MIBK	AAS	[179]
$(NH_4)_2 HPO_4$	Ca	Oxine	3-Methyl-1-butanol	AAS	[180]
Al	Ga	8-Quinolinethiol	MIBK	Fluor.	[181]
U compounds	Ag, Cu, Hg	Dithizone	Carbon tetrachloride	Phot.	[182]
W and its oxide	Co, Cu, Ni, Pb, Zn	Dithizone	Chloroform	XRF	[183]
Se	Ag, Al, Au, Bi, Cd, Co, Cu, Fe, Ga, Hf, Hg, In, La, Mn, Mo, Ni, Pb, Pd, Pt, Sb, Sc, Sn, Th, Ti, Tl, U, V, Y, Zn, Zr	Dithizone + oxine	Chloroform	OES	[184]
Ge, Si and their compounds	As, Bi	Diethylammonium diethyldithiocarbamate	Chloroform	Gutzeit test, Phot., OES	[185–188]
Al, Cu and their alloys	Ag	DDTC	Benzene	AAS	[189]
TiO_2	V	DDTC	Chloroform	AAS	[190]
Glasses, Na_2CO_3, $CaCO_3$	Co, Cr, Cu, Fe, Mn, Ni	DDTC	MIBK	AAS	[191]
Rocks	Co, Cu, Ni	DDTC	MIBK	AAS	[192]
Al, Ti, Zr and their compounds	Ag, As, Au, Bi, Cd, Co, Cr, Cu, Fe, Ga, Hg, In, Mn, Mo, Ni, Pb, Pd, Pt, Sb, Se, Sn, Te, Tl, U, V, Zn	APDC + dithizone	Chloroform	OES	[193]
Be and its oxide	Li	Dipivaloylmethane	Ethyl ether	Phot.	[194]

5.3 Extraction of Ion Pairs

5.3.1 Ion-Association Systems

Uncharged compounds formed by the association of oppositely charged ions in pairs can be extracted from aqueous solutions into organic solvents. The coulombic force of attraction, which contributes to the formation of ion pairs, is proportional to

$$z_+ z_- / \left\{ \epsilon (r_+ + r_-)^2 \right\} \ ,$$

where z and r are the charges and radii of the ions, respectively, and ϵ is the dielectric constant of the solvent. The anions or cations are frequently complex ions, and solvation plays an important role in the ion-association extractions with organic solvents such as ethers and TBP.

Typical ion-association extraction systems are shown in Table 19.

Table 19. Typical ion-association extraction systems [152]

Aqueous phase	Organic phase	Elements extracted
6 M HCl	Ethyl ether	As, Au, Fe, Ga, Ge, Mo, Sb, Tl (E = 51−100%); Sn, Te (E = 6−50%); Hg, Ir, Zn (E = 0.1−5%)
4.5−5 M HBr	Ethyl ether	As, Au, Fe, Ga, In, Sb, Sn, Tl (E = 51−100%); Mo, Se (E = 6−50%); Cd, Cu, Hg, Te, Zn (E = 0.1−5%)
6.9 M HI	Ethyl ether	As, Au, Cd, Hg, Sb, Sn, Tl (E = 51−100%); Bi, In, Mo, Te, Zn (E = 6−50%)
20 M HF	Ethyl ether	Nb, Re, Ta (E = 51−100%); As, Ge, Mo, P, Sb, Se, Te, V (E = 6−50%); Al, Be, Cd, Co, Cu, Hg, Mn, Ni, Sn, U, Zn, Zr (E = 0.1−5%)
8 M HNO$_3$	Ethyl ether	Au, Ce, Np, Tc, U (E = 51−100%); As, Bi, Cr, P, Th, Tl, Zr (E = 6−50%); Ag, Be, Cd, Co, Cu, Fe, Ga, Ge, Hg, In, La, Mn, Mo, Pb, Sb, Sc, Ti, V, Y, Zn (E = 0.1−5%)
SCN$^-$	Ethyl ether	Au, Be, Co, Fe, Ga, In, Mo, Nb, Os, Re, Rh, Ru, Sc, Sn, Ti, W, Zn (E = 51−100%); Al, Hf, U, V, Zr (E = 6−50%); Ag, As, Bi, Cd, Cr, Cu, Ge, Hg, Pd, Sb (E = 0.1−5%)
6 M HCl	TBP	As, Au, Cd, Co, Fe, Ga, Ge, Hg, In, Mo, Nb, Np, Os, Pa, Pd, Pt, Re, Sb, Sc, Sn, Ta, Tc, Te, Tl, U, V, W, Zn, Zr (D > 1)
6 M HNO$_3$	TBP	Au, Hf, Nb, Np, Os, Pa, Pu, Sc, Th, U, Y, Zr (D > 1)
8 M HCl	Tri-isooctylamine (a high-molecular-weight amine) in xylene	Au, Bi, Cd, Co, Cu, Fe, Ga, In, Mo, Nb, Np, Os, Pa, Pu, Re, Ru, Se, Tc, Te, U, V, W, Zr (D > 1)

The theoretical treatment of extraction equilibria in ion-association systems is more difficult than that of chelate extraction systems, because of the following two reasons.

First, there exist a relatively *large number of chemical species and equilibria* which may participate in the extraction. For example, in the extraction of Fe(III) from hydrochloric acid solutions into ether, we have to consider:

(1) Dissociation of reagent, hydrogen chloride, in the aqueous phase,

(2) Stepwise formation of iron(III) chloride complexes in the aqueous phase, i.e. $FeCl^{2+}$, $FeCl_2^+$, $FeCl_3$ and $FeCl_4^-$,

(3) Formation of extractable complex and reagent, i.e. solvated ion pairs (H^+, $FeCl_4^-$) and (H^+, Cl^-), in the aqueous phase,

(4) Distribution of the extractable complex and reagent between the aqueous and organic phases,

(5) Polymerization of the complex in the organic phase having a low dielectric constant, and

(6) Dissociation of the extractable complex and reagent in the organic phase having a high dielectric constant.

Second, there are *great changes in activity coefficients, dielectric constants, and volumes* of the aqueous and organic phases, due to high electrolyte concentrations generally used in ion-association extraction systems and the resulting enhanced mutual solubility of the two phases. Corrections of these factors are difficult or impossible, and the utility of the mass action expressions used in describing the extraction equilibria is greatly reduced.

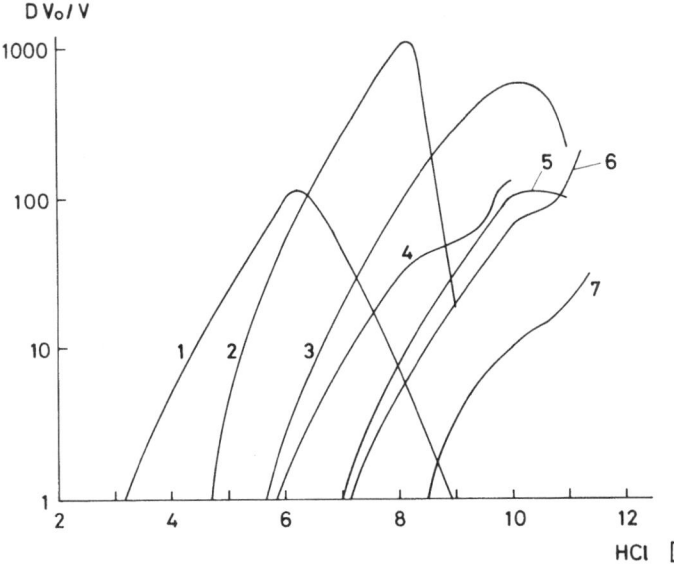

Fig. 19. Extraction of iron(III) with ethers from hydrochloric acid solutions [201]. 1: Ethyl ether, 2: Isopropyl ether, 3: Propyl ether, 4: Bis(2-chloroethyl)ether, 5: Butyl ether, 6: Pentyl ether, 7: Isopentyl ether. $V_0/V = 1$ (initial)

Table 20. Ion-association extraction of trace elements

Matrices	Trace elements	Aqueous phase	Organic phase	Determination techniques	Lit.
Cd	Tl	8 M HCl	Ethyl ether	Polar.	[202]
In	Fe, Ga, Tl	6 M HCl	Ethyl ether, Isopropyl ether	Polar., Phot.	[203–205]
Al	Fe, Ga	HCl	MIBK, Ethyl ether	Phot.	[206, 207]
Se	Au	HCl, HCl-HBr	Methyl methacrylate	AAS	[208]
Cu	Au	HCl-HNO$_3$	Ethyl acetate	Phot.	[209]
Sb	As	10–12 M HCl	Benzene	AAS	[210]
Al, Fe- and Ni-base alloys	Ag, Bi, Pb, Sb, Sn	HCl + KI	Tri-n-octylphosphine oxide in MIBK	AAS	[211, 212]
Ni	Co, Fe	SCN$^-$	TBP	Phot.	[213, 214]
Bi	Zn	SCN$^-$	Ethyl ether	Polar.	[215]
U	B	Tetraphenylarsonium chloride soln.	Chloroform	OES	[216]
Th compounds	Cd	HCl	High-molecular-weight amine in xylene	Polar.	[217]
Zn	Cd	KI soln. (pH 3)	High-molecular-weight amine in xylene	Phot.	[218]
Al, Cu, Ni and other nonferrous metals and alloys	Au, Bi, Cd, Pb, Pd, Sn, Te, Zn	HCl, HBr	High-molecular-weight amines (or quaternary ammonium salts) in MIBK (or butyl acetate)	AAS	[219–223]
Underground water	Al, Ca, Co, Cr, Cu, Eu, Fe, La, Mg, Mn, Sc, Sm, Zn	pH 2	Dinonylnaphthalene sulfonic acid in hexane	NAA	[224]
River water	Cd	KI + zephiramine	Ethyl acetate	AAS	[225]

Table 21. Ion-association extraction of matrix elements

Matrices	Aqueous phase	Organic phase	Trace elements	Determination techniques	Lit.
Fe	HCl	Ethyl ether, MIBK, Isobutyl acetate	Ag, Al, As, Bi, Co, Cr, Cu, Mn, Ni, Pb, Ti, V, Zr	OES, Polar.	[226–228]
Ga	6–8 M HCl	Ethyl ether, Isopropyl ether	Al, Be, Bi, Ca, Cd, Co, Cr, Cu, Mg, Mn, Mo, Ni, Pb, Ti, V, Zn, Zr	OES, Phot., AAS	[229–232]
GaAs*	8 M HCl	Isopropyl ether	Al, Be, Bi, Cd, Co, Cr, Mg, Mn, Ni, Si, V, Zr	OES	[94]
Te	HCl-HNO$_3$	MIBK	Cu, Pb	Polar.	[233]
Au	HCl	Ethyl ether, Ethyl acetate, MIBK	Ag, As, Bi, Cd, Cu, Fe, Mg, Mn, Ni, Pb, Pd, Pt, Rh, Sb	Polar., Phot., OES	[234–236]
Au	3 M HBr	Isopropyl ether	Bi, Cd, Cu, Fe, Ni, Pb, Pd, Pt, Zn	Polar., Phot.	[237, 238]
In	5 M HBr	Isopropyl ether	Al, Ba, Be, Bi, Ca, Cd, Co, Cr, Cs, Cu, Eu, K, Li, Mg, Mn, Na, Ni, Pb, Pd, Pt, Rb, Rh, Sc, Sr, Ti, V, Y, Yb, Zn, Zr	Polar., OES, Phot.	[203, 239, 240]
Hg	HI or HI-HBr	Cyclohexanone	Bi, Cd, Cu, Fe, Mn, Ni, Pb, Zn	Phot., Voltam.	[241]
Bi	HI	Cyclohexanone	As, Cu, Fe, Ni, Pb, Zn	Phot., Polar., Voltam.	[242]
Cd	HI	TBP	Cu, Fe, Ni, Pb, Zn	Polar., Phot.	[243]
U	HCl	TBP-chloroform	Mn	Phot.	[244]
CeO$_2$	HNO$_3$	TBP	Ag, Al, Bi, Ca, Co, Cu, Fe, Mg, Mn, Ni, Pb, Sn, Ti	OES	[245]
U	3 M HNO$_3$	Dioctyl sulfoxide in xylene	Cd, Co, Cu, Ni	AAS	[246]
Ag	HNO$_3$	O-Isopropyl-N-methyl-thiocarbamate in chloroform	As, Bi, Cd, Co, Cr, Cu, Fe, Ir, Mn, Mo, Ni, Pb, Rh, Sb, Se, Te, Zn	AAS	[247]

* After removal of As by volatilization

The complicated dependence of the distribution ratio of Fe(III) on acidity in the hydrochloric acid-ether extraction systems is shown in Fig. 19. The distribution ratio generally depends on the iron concentration.

The enhancement of the extractability in ion-association systems is effected by the addition of electrolytes called salting-out agents to the aqueous phase. For example, in the extraction of U(VI) from aqueous nitrate solutions into ethyl ether, addition of large amounts of nitrates of alkali and alkaline earth elements markedly increases the distribution ratio. The salting-out effect may be attributed to the increase in anion concentrations as well as the decreases in the activity of water and the dielectric constant of the aqueous phase.

5.3.2 Ion-Association Extraction of Trace Elements

Table 20 tabulates some applications of this technique in the analysis of high-purity metals and other samples for impurities at the low μg/g or the ng/g level.

5.3.3 Ion-Association Extraction of Matrix Elements

Table 21 lists typical examples of the extraction of matrix elements into the organic phase, leaving the desired trace elements in the aqueous phase, for the enrichment of impurities at the low μg/g or the ng/g level in metals and their compounds.

5.4 Special Extractions

5.4.1 Three-Phase Extraction

In some extraction systems, two organic phases and one aqueous phase are obtained after equilibrium. For example, when diantipyrylmethane in a 7 : 3 mixture of benzene and chloroform is equilibrated with an aqueous solution containing various metal ions, the following three phases appear:
the aqueous phase as top layer,
the main organic phase as middle layer, and
the other organic phase of small volume as bottom layer which contains 95 to 98% of the metal elements [248].

The third phase is transferred directly to a carbon electrode for the emission spectrographic determination of such elements as Bi, Cd, Hf, Hg, Sc, Sn and Zr. Time-consuming evaporation and ashing procedures are dispensed with.

5.4.2 Homogeneous Extraction

Some organic solvents are immiscible with water at room temperature, but become miscible at higher temperatures. This property is successfully applied to the extraction of iron(III) with TTA, which needs a very long time to establish equilibrium in the conventional extraction. An aqueous solution containing iron(III) and a propylene carbonate solution of TTA are gently shaken to form a homogeneous solution at

80 °C, where the reaction rate of the iron(III)-TTA complex formation is rapid. The solution is then cooled to room temperature and centrifuged to form two separate layers [249]. In another example of homogeneous extraction, an iron(III)-TTA complex is formed in a homogeneous solution, one-to-one mixture of 2-propanol and water, and the solution is separated into two phases by adding sodium nitrate as salting-out agent. During the above process, the iron(III) is quantitatively extracted into the organic phase [250]. The extracted species may be ion pairs in both cases.

5.4.3 Extraction with Molten Organic Compounds

Various metal chelates are extracted from aqueous solutions into molten organic compounds such as naphthalene, biphenyl and stearyl alcohol at elevated temperatures [251−255]. After cooling, the solid phase is separated from the liquid phase and delivered to spectrophotometry or polarography after dissolving in an appropriate organic solvent or to X-ray fluorescence spectrometry after pressing in a die to form a thin film.

5.4.4 Extraction of Trace Elements from Nonaqueous Samples

Copper, Fe, Ni, Pb and V at the ng/g level in petroleum distillates are extracted with sulfuric acid and then with a hydrochloric acid-acetone-water mixture at 100 °C, and determined by emission spectrography [256, 257]. Sodium in liquid hydrocarbons is extracted with water and determined by flame photometry [258]. Copper and lead at the low ng/g level in catalytic reformer feedstocks are extracted with 4 M hydrochloric acid for the polarographic determination [259]. Phosphorus and boron at the ng/g or low μg/g level in silicon tetrachloride are extracted with concentrated sulfuric acid and quinalizarin-sulfuric acid, respectively, and determined by spectrophotometry [260, 261].

Fire assay and some other techniques using a liquid metal and the other liquid phase are regarded as liquid-liquid extraction in a broad sense (see Chaps. 6 and 8).

6 Selective Dissolution

Either the matrix or desired trace elements in a solid (or liquid metal) sample are selectively dissolved in an appropriate solvent. This technique is also called solid-liquid extraction when the sample is solid.

6.1 Selective Dissolution of the Matrix

This technique is widely used in enrichment of various metal oxides, carbides and nitrides existing as inclusions in or surface films on steels and nonferrous metals and alloys [53]. Some examples are given in Table 22. The sample is simply decomposed with a suitable inorganic or organic solvent, in which the matrix is soluble but oxides etc. are not, in a flask with a reflux condenser. Heating, stirring and ultrasonic irradiation are applied to accelerate the dissolution. After the dissolution, the residue is separated from the solution by filtration, further separated by selective dissolution, magnetic separation and sieving, if necessary, and delivered to optical and electron microscopic observation, X-ray diffraction or elemental analysis. This technique is

Table 22. Selective dissolution of metal matrices for separating inclusions and surface films as residues

Matrices	Solvents	Residues	Lit.
Steels	Iodine in methanol, 10 M H_3PO_4, 6 M HCl	Oxides, Carbides, Nitrides	[146, 262, 263]
Copper-base alloys	Bromine + methanol, 7 M HNO_3	Oxides	[264, 265]
Nickel-titanium alloy	Bromine + methanol, Bromine + methyl acetate	Oxide, Carbide, Nitride	[266]
Al	Bromine + methanol, Phenol, 4 M HCl, HCl-HNO_3	Oxide	[267–269]
Mg	Phenol	Oxide	[270]
Ni	14% $KCuCl_3$	Oxide	[271]
Bi	Hg*	Oxide	[272]

* Shaken together with the sample and a 1 M NH_4NO_3 solution in a separatory funnel to dissolve the matrix in the mercury. The residue remains on the mercury-solution interface.

Table 23. Partial dissolution of metal matrices for concentrating trace impurities in residues

Matrices	Solvents	Trace elements enriched	Lit.
Hg*	10 M HNO$_3$	Au, Pd	[273]
Ga*	3 M HCl-0.15 M HNO$_3$	Ag, Au, Bi, Co, Cu, Fe, Hg, Ni, Pb, Pd, Sn	[274, 275]
Pb	2.5 M HNO$_3$	Ag, Au, Pd	[276]
Zn	12 M HCl	Ag, Au, Bi, Cd, Co, In, Ni, Pb, Pd, Sn, Tl	[277]
Cd	5% HNO$_3$	Ag, Au, Bi, Cu, Ni, Pb, Pd	[278]
	40% HBr	Ag, Au, Bi, Co, Cu, Fe, In, Ni, Pb, Pd, Sn	[279]
Al	1.5 and 6 M HCl	Bi, Cd, Ga, In, Pb, Tl, Zn	[280]
Mn	3 M HCl	Ag, Au, Bi, Cd, Co, Cu, Fe, Ga, In, Ni, Pb, Pd, Tl	[281]
In	47% HBr	Ag, Au, Bi, Co, Cu, Fe, Ni, Pb, Pd	[282]
Sn	HBr-Br$_2$	Ag, Au, Bi, Cu, Pd	[282]

* liquid metals

applicable to the μg/g level, if care is taken to prevent losses due to partial dissolution of the desired constituents and incomplete collection of the residues.

Partial dissolution of the matrix is useful for multielement enrichment of trace impurities, which are electrochemically nobler than the matrix, in high-purity metals. Applications of this technique are summarized in Table 23. Prior to dissolution of solid metals, a sample is first coated with a thin layer of mercury, through which the matrix is transported and wherein the trace impurities remain quantitatively during the dissolution step. A metal sample weighing 5 to 100 g is dissolved in a solvent until 5 to 100 mg of the matrix element remains. The residue is then separated from the solution by decantation and analyzed for trace elements by atomic absorption spectrometry, spectrophotometry, polarography and emission spectrography, after removal of, or in the presence of, mercury. Recoveries for trace impurities at the ng/g or low μg/g level are better than 95%, with enrichment factors of 10^2 to 10^3.

6.2 Selective Dissolution of Trace Elements

In this case, it is essential that desired trace constituents in a sample are brought into thorough contact with a solvent, in which the desired trace constituents are soluble but the matrix is not.

From this viewpoint, selective dissolution of trace impurities in liquid metals or on the surfaces of solid samples is rather easy. Such impurities as Cd, Co, Fe, Tl and Zn at the ng/g or low μg/g level in high-purity gallium are extracted with hot 0.2 M HI-0.05 M I$_2$ without dissolving the matrix and determined by atomic absorption spectrometry [275]. Oxides on high-purity metal surfaces are selectively dissolved in the following solutions in a nitrogen atmosphere: 1 M ammonium chloride-ammonium carbonate-aqueous ammonia (pH 10) for copper [283], 1 M ammonium chloride-aqueous ammonia (pH 10) for cadmium [284], and 0.5% ammonium acetate with sodium borohydride as deoxygenating agent for lead [285]. The dissolution is effected by shaking for 5 min in a separatory funnel. In the presence of metallic mercury, the cadmium or lead metal matrix is simultaneously amalgamated (dissolved in mercury), and the total oxides on the surface and in the interior of the samples are dissolved in the above solutions. The metal ions in the solutions are determined by atomic absorption spectrometry and the quantity of the corresponding oxide is calculated.

In other applications of this technique to trace impurities in metals, the *metal matrix is converted into another compound* during or before the separation step as shown in Table 24. For example, two methods are proposed for the enrichment of B at the ng/g level in silicon. In Luke and Flaschen's method, silicon is converted into crystalline silica by heating the sample with a small amount of 0.5% sodium hydroxide solution in an autoclave at 350 °C and 350 atm. Boron present in the liquid phase is then determined by spectrophotometry or fluorometry [286, 287]. In Morrison and Rupp's method, the sample is dissolved in a sodium hydroxide solution, and the sodium ion in the solution is removed by electrolysis through a cation-permeable membrane until the pH of the solution is lowered to between 7 and 8 (Fig. 20). The

Table 24. Selective dissolution of trace elements

Matrices	Converted into	Solvents	Trace elements dissolved	Lit.
Si	Oxide	0.5% NaOH, Water	B	[286–289]
Te	Oxide	Water, 0.1 M HNO_3, 0.1 M HCl	Cd, Co, Zn	[290]
Na	Chloride	95% Ethanol	B	[291]
	Chloride	Ethanol, 1-Butanol-12 M HCl (1 : 1), Acetone	Co, Cu, Fe, Zn	[292]
K	Chloride	Ethanol, 1-Butanol-12 M HCl (9 : 1), Acetone	Co, Cu, Fe, Zn	[292]
Ba	Chloride	12 M HCl, 1-Butanol-12 M HCl (1 : 1)	Co, Fe, Zn	[292]
Ni	Chloride	Acetone-12 M HCl (99 : 1), (9 : 1)	Co, Cu, Fe, Zn	[292]
Cd	Chloride	Acetone-12 M HCl (999 : 1)	Co, Cu, Fe, Zn	[292]
Pb	Chloride	Ethanol-12 M HCl (999 : 1), Acetone-12 M HCl (999 : 1)	Cd, Co, Cu, Fe, Zn	[292]
	Nitrate	14 M HNO_3, Ethanol-12 M HCl (99 : 1)	Ag, Co, Fe, Zn	[293]
	Sulfate	Ethanol-14 M HNO_3 (3 : 1)	Co, Fe, Zn	[293]
Bi	Basic nitrate	0.07 M H_3PO_4- 0.12 M HCl, Water	Co, Cu, Fe, Zn	[294–296]

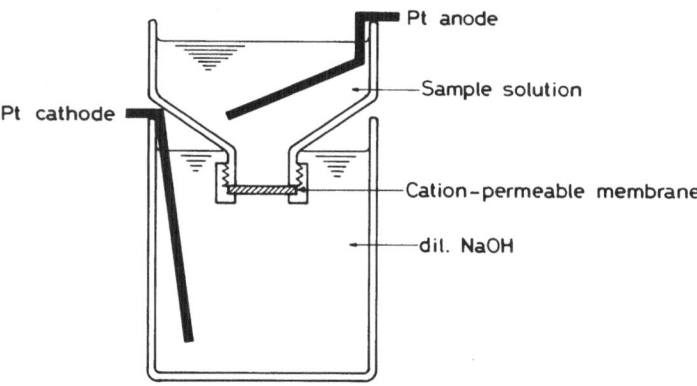

Fig. 20. Polyethylene electrolysis cell (After Morrison and Rupp [288])

solution is then evaporated, and B is extracted from the residue with water and determined by emission spectrography or isotope dilution-mass spectrometry [288, 289]. Many other metals are decomposed with mineral acids, and the resulting solutions are evaporated to dryness to redistribute the trace elements on the surfaces or in the interstitial space of the agglomerates of the pure matrix crystals (oxides and salts). Sometimes, the solids are further heated at higher temperatures and converted into other compounds to reduce the solubility of the matrix [296]. Then the trace elements, together with a small amount of the matrix, are quantitatively dissolved in water, mineral acids, organic solvents or mixtures of the above solvents under ultrasonic irradiation. For impurities at the ng/g or low μg/g level, recoveries greater than 95% are obtained with enrichment factors of 10^2 to 10^4 by this technique.

In *fire* or *dry assay* for silver, gold and the platinum metals in ores and concentrates, a mixture of sample, lead (or copper) oxide, flux and reducing agent is heated in a crucible to high temperatures [297–299]. The noble metals are selectively dissolved in the molten lead (or copper) that is formed, leaving nonvaluable minerals or gangue as liquid slag. The lead (or copper) button at the bottom of the cooled crucible is separated from the slag, and after removal of lead (or copper), the noble metals are determined.

It is common practice in agricultural and ecological studies to extract trace elements in the *soil* with solutions of acid, salts or chelating agents, and boiling water. Such soluble fractions are supposed to be plant-available [300].

7 Precipitation

Precipitation [301–303], one of the oldest methods of separation, is still very useful as an enrichment technique in inorganic trace analysis.

7.1 Precipitation of Matrix Elements

Under proper conditions, the matrix element can be removed by precipitation, leaving the desired trace elements quantitatively in an aqueous sample solution, by using one of the following methods:

(1) Addition of a liquid or gaseous precipitant that reacts with the matrix element to form a sparingly soluble compound. This is the most conventional technique. Although the precipitant is generally added in excess to reduce the concentration of the matrix element in the solution as low as possible, its large excess should be avoided when the soluble complexes of the matrix element are formed or when the desired trace elements also react with the same precipitant to form the precipitates.

(2) Change of the pH of the solution in the presence of the precipitant.

(3) *In situ* slow generation of the precipitant or other species which participate in the precipitation. For example, hydroxyl ions are generated by heating an aqueous sample solution containing urea. Urea has negligible basic properties, but its hydrolysis proceeds at a controlled rate upon heating at appropriate temperatures, raising the pH of the solution.

$$(NH_2)_2 CO + H_2O \rightarrow 2NH_3 + CO_2$$

In this kinetically controlled precipitation technique, the concentration of the precipitant is always kept uniform and low throughout the solution. The degree of supersaturation is lower and a smaller number of crystal nuclei are formed compared with the external addition of the precipitant or other reagents. Thus larger and more perfectly shaped crystals are formed, which are advantageous because of easier filtration and lower incorporation of impurities. This technique called precipitation from homogeneous solution [304] is applicable to many precipitations as shown in Table 25.

(4) Change of solvent to reduce the solubility of the matrix element. This can be effected, for example, by saturation with hydrogen chloride or by addition of concentrated mineral acids or water-miscible organic solvents.

(5) Removal of solvent by evaporation.

Table 25. Precipitation from homogeneous solution

Precipitation as	Reagents used
Hydroxide and basic salt	Urea, Acetamide
Phosphate	Trimethyl or triethyl phosphate, Metaphosphoric acid
Oxalate	Dimethyl or diethyl oxalate, Urea + oxalic acid
Sulfate	Dimethyl or diethyl sulfate, Sulfamic acid
Sulfide	Thioacetamide
Iodate	Periodate + β-hydroxyethyl acetate, Periodate + ethylene diacetate
Carbonate	Trichloroacetate
Chromate	Urea + potassium dichromate
Chelate	Urea + chelating agents, Organic compounds generating chelating agents by hydrolysis or other reactions

Generally, milligram quantities of the matrix element remain in the solution in the last two techniques.

After completion of the precipitation, the precipitates are separated by filtration or centrifugation, and washed thoroughly with appropriate solutions.

The lower the solubility of the precipitate of the matrix element, the higher the enrichment factor can be attained. With regard to the solubility of an ionic precipitate, we must generally consider, in addition to the ionic solubility calculated by the solubility product constant of the precipitate, the molecular or intrinsic solubility, i.e. solubility of undissociated molecules, and existance of soluble complex ions formed by addition of an excess of the precipitant. The latter two can not be neglected in some cases.

Matrix precipitation often enables the simultaneous multielement enrichment of the desired trace elements. Generally speaking, however, this technique has three disadvantages compared with carrier precipitation of trace elements; contamination due to larger amounts of precipitant used, possible losses due to coprecipitation phenomena discussed below, and larger final solution volumes.

7.1.1 Coprecipitation Phenomena

Coprecipitation is defined as the contamination of a precipitate by substances that normally remain in the solution under the conditions of the precipitation. The following three mechanisms are considered to be responsible for coprecipitation phenomena.

Mixed crystal formation. A matrix ion M in the ionic crystal lattice of the matrix precipitate MR can be replaced by a trace ion T of the same sign to form a mixed crystal (M, T)R, when the compounds MR and TR are isomorphous and their lattice constants (or ionic radii of M and N) do not differ too much from each other. Solid compounds of dissimilar crystallographic types or lattice constants can also form mixed crystals having a limited miscibility. The above two types are often called the isomorphous and the anomalous mixed crystal formations, respectively. When there is a difference in the charge of M and T, it is compensated by lattice vacancies or by

simultaneous substitutions of the sites of the opposite charge (e.g. $KMnO_4$ in $BaSO_4$).
Mixed crystal formation is responsible, for example, for the coprecipitation of traces
of lead with barium sulfate under the conditions where the ion product $[Pb^{2+}]\,[SO_4^{2-}]$
is below the solubility product constant of $PbSO_4$.

When the crystal is completely homogeneous with respect to the trace element
after the sufficient aging in the solution, the distributions of the trace and matrix
elements between the liquid and solid phases may be expressed by the Berthelot-
Nernst equation:

$$\frac{Q_T}{Q_M} = D \; \frac{Q_T^0 - Q_T}{Q_M^0 - Q_M} \tag{32}$$

where Q_T^0 and Q_M^0 are the total quantities of the trace and matrix elements in the
system, respectively, Q_T and Q_M are the quantities of the trace and matrix elements
in the precipitate, respectively, and D is the homogeneous distribution coefficient.
Ideally, D is equal to the ratio of the solubility product constants of the matrix and
trace elements for 1 : 1 type crystals. The homogeneous distributions for various values
of D are shown in Fig. 21.

If equilibrium exists between the solution and an infinitesimally thin surface layer
of the crystal,

$$\frac{dQ_T}{dQ_M} = \lambda \; \frac{Q_T^0 - Q_T}{Q_M^0 - Q_M} \tag{33}$$

where λ is the logarithmic distribution coefficient. By integration, we obtain the
Doerner-Hoskins equation:

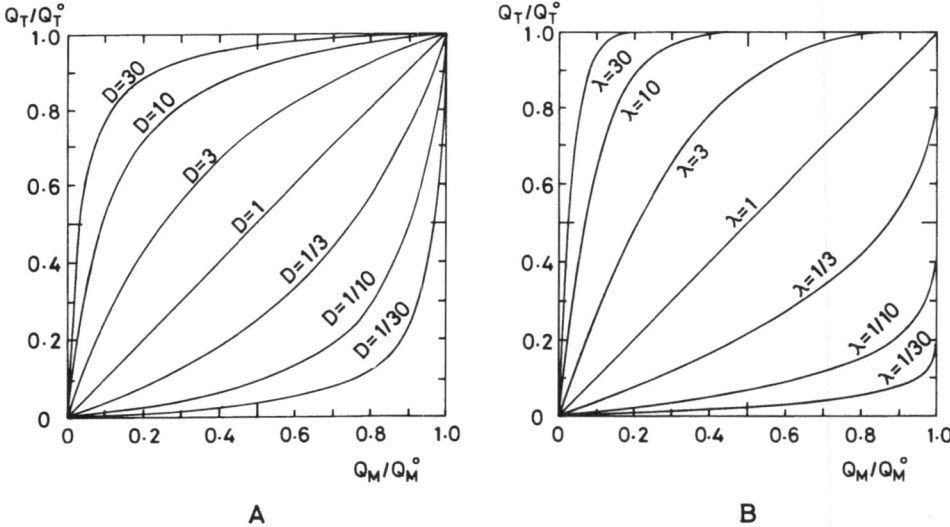

Fig. 21. Homogeneous (A) and logarithmic (B) distributions

$$\log (1 - Q_T/Q_T^0) = \lambda \log (1 - Q_M/Q_M^0) \tag{34}$$

Ideally, λ is equal to D of Eq. (32). The λ is dependent on the rate of precipitation, and if greater than unity, it decreases with increasing precipitation rate. The logarithmic distributions for various values of λ are shown in Fig. 21. Equation (34) is obeyed, for example, when a trace of radium is coprecipitated by slow evaporation of a saturated solution of barium bromide at a constant temperature, homogeneous precipitation with barium chromate, or rapid precipitation of barium nitrate from supersaturated solutions by vigorous agitation for a few minutes at 0 °C, and then the precipitates are filtered off immediately.

Usually, actual coprecipitations may fall somewhere between the above two limiting distributions.

Surface adsorption. Unlike ions within an ionic crystal, those at the surface of the crystal are not completely surrounded by their neighbor ions of opposite charge, and, hence, attract other ions of opposite charge from the solution, especially ions common to those within the crystal. For example, barium sulfate precipitates strongly attract barium and sulfate ions in the solution to their surfaces. If only one of the common ions is present in the solution, impurity ions of opposite charge are also attracted to maintain electrical neutrality. Traces of lead sulfate are thus adsorbed on the surfaces of calcium sulfate crystals.

Adsorption due to ion exchange also occurs on the crystal surfaces. For example, traces of lead ions are adsorbed on the surfaces of barium sulfate, even in the absence of sulfate ions in the solution, by the following ion exchange reaction:

$$BaSO_4(solid) + Pb^{2+} \rightleftharpoons PbSO_4 (solid) + Ba^{2+}$$

Similar phenomena take place during the formation of colloidal or amorphous precipitates.

According to the Fajans-Paneth-Hahn rules: (1) the adsorption of impurity ions will be large, if the surface area of the crystal is large, and if the impurity ions form sparingly soluble compounds with one of the lattice-ions of the crystal, and (2) the adsorption of cations is increased by a negative charge on the surface due to adsorbed anions, and decreased by a positive charge on the surface due to adsorbed cations, and vice versa.

Occlusion. Occlusion of trace ions may occur when the trace ions are adsorbed on the surface of a growing precipitate and covered by subsequent layers of the precipitate. Sometimes, the solvent, which contains the trace ions, is mechanically trapped within the precipitate. Occlusion is more prevalent with colloidal precipitates than with large-crystalline precipitates, and in rapid precipitation than in slow precipitation.

Because actual coprecipitation phenomena are very complicated and their extent varies greatly with precipitation conditions as well as concentration levels of trace elements, it is always necessary to confirm trace recoveries experimentally.

To minimize losses of the desired trace elements due to coprecipitation with the matrix precipitates, the following means may be useful:

(1) Masking of the desired trace elements with a suitable complexing agent (see Appendix A. 2).

(2) Precipitation from homogeneous solutions or other slow precipitation techniques.

(3) Prolonged digestion or aging of the matrix precipitate to effect recrystallization.

(4) Reprecipitation.

(5) Washing of the precipitate with an appropriate solution.

Formerly, the matrix precipitation was unpopular in inorganic trace analysis, because the hazard of losses of desired trace elements due to coprecipitation was considered to be very large. A number of recent investigations have revealed the unfair underestimation of this technique. Under proper conditions, coprecipitation losses are negligible for numerous trace elements.

7.1.2 Application of Matrix Precipitation

Table 26 lists some applications of matrix precipitation to the analysis of high-purity metals and compounds for trace impurities at the ng/g or low μg/g level.

Jackwerth [305] compared the following three methods for multielement preconcentration from pure lead by precipitation of the matrix:

(1) *Precipitation as nitrate:* A 10-g sample is dissolved in 75 ml of 20% nitric acid, and the solution is evaporated at temperatures slightly below the boiling point to a

Table 26. Applications of matrix precipitation

Matrices	Precipitated as	Trace elements	Determination techniques	Lit.
Pb	$PbNO_3$	Ag, Al, Bi, Cd, Co, Cu, Fe, Ga, In, K, Mg, Mn, Na, Ni, Pd, Tl, Zn	AAS	[305]
Pb	$PbCl_2$	Ag, Al, Au, Bi, Cd, Co, Cu, Fe, Ga, In, K, Mg, Mn, Na, Ni, Sb, Tl	AAS, Phot.	[305, 306]
Pb	$PbSO_4$	Al, Cd, Co, Cu, Ga, In, Mn, Ni, Pd, Zn	AAS	[305]
Pb	PbS_2O_3	Zn	Phot.	[307]
Tl	TlI	Bi, Cd, Co, Cu, Fe, In, Ni, Pb	OES, Phot.	[308]
Hg	Hg	Bi, Cd, Co, Cu, Fe, Mg, Mn, Ni, Pb, Tl, Zn	Voltam., OES, Phot.	[309]
Ag	Ag amalgam	As, Cd, Cu, Fe, Ga, In, Mn, Ni, Pb, Tl, Zn	Voltam., OES, Phot.	[310]
Ni	Hexammine nickel perchlorate	Co	Phot.	[311]
Te	TeO_2	Cu, Pb	Phot., Polar.	[312, 313]
Si, Ge	Sodium silicate or germanate	B	Phot.	[314]
Cu	CuSCN	Fe, Pb	Phot.	[315]
Cu	CuS	Cd, Co, Fe, In, Mn, Ni, Pb, Zn	AAS	[316]

moist crystal conglomerate. This is then treated with slight heating with 25 ml of 65% nitric acid, and after cooling by running water, the solution is decanted. The above treatment is repeated twice more with 10 ml each of 65% nitric acid without heating. The combined solutions are evaporated to near dryness, and the residue is dissolved in warm 1 M nitric acid for trace determinations.

(2) *Precipitation as chloride:* A 10-g sample is dissolved in 40 ml of 20% nitric acid. To the solution, 30 ml of 37% hydrochloric acid is added, and the resulting suspension is evaporated at temperatures slightly below the boiling point to half of its initial volume. After addition of 50 ml of water and cooling by running water, the suspension is centrifuged. The supernatant solution is decanted and evaporated to near dryness. The residue is treated with 10 ml of 1 M hydrochloric acid for an hour (overnight for antimony recovery). After centrifugation, the supernatant solution is used for trace determinations.

(3) *Precipitation as sulfate:* A 10-g sample is dissolved in 40 ml of 20% nitric acid followed by 50 ml of water. To the vigorously stirred hot solution, 8 ml of 33% sulfuric acid is added, and after cooling by running water, the suspension is centrifuged. The solution is decanted and evaporated to dryness. The residue is treated with 5 ml of 1 M hydrochloric acid with slight heating, and the supernatant solution is used for trace determinations.

Impurity elements enriched in greater than 90% yields by the above three procedures are shown in Table 26. Enrichment factors of about 10^3 are obtained in all the procedures. In general, precipitation as nitrate is superior to the other two from the viewpoints of operational simplicity and rapidity as well as trace recoveries. Precipitation as chloride is recommended only for the enrichment of gold and antimony.

7.2 Precipitation of Trace Elements

When the desired trace elements are present in an aqueous sample solution at concentrations below about 1 mg/l, it is generally difficult or impossible to precipitate and separate them quantitatively by conventional precipitation techniques. Even if they have very small solubilities or solubility product constants, supersaturation of the solution, formation of colloids or minute amounts of the precipitates may cause difficulties. Therefore, carrier precipitation is generally applied to ensure quantitative trace recoveries. Instead of the conventional precipitation in aqueous solutions, trace elements can also be precipitated at the interface between an aqueous solution and an organic solvent by electrolysis [317] or by shaking the two phases in the presence of basic dyes [318].

7.2.1 Carrier Precipitation

Carrier precipitation is defined as collection of the desired trace elements in solutions on a milligram quantity of another precipitate, which is called *a collector, carrier* or *gathering precipitate*, by coprecipitation and the simple mechanical carrier action. In this case, coprecipitation, a disadvantage in matrix precipitation, is successfully taken advantage of. Collector precipitates are formed in aqueous sample solutions by one of the following methods:

(1) Addition of a carrier element followed by an inorganic or organic precipitant to form a collector precipitate.

(2) Precipitation of an element originally present in milligram quantities in the sample solution.

(3) Partial precipitation of a matrix element having a larger solubility product constant than those of trace elements. Examples include the sulfide precipitation of traces of Cd in the zinc matrix and Pb in the iron matrix.

(4) Addition of a water-insoluble organic precipitant (e.g. *p*-dimethylamino-benzylidenerhodanine, dithizone) dissolved in a water-miscible organic solvent (e.g. alcohols, acetone, methyl cellosolve) to deposit the precipitant itself.

It must be kept in mind that matrix elements may sometimes disturb the formation of collector precipitates at proper pH or lower the trace recoveries.

After the collection of the desired trace elements, the collector precipitates are separated from the sample solution by filtration or centrifugation, followed by washing with small amounts of water or appropriate solutions, if necessary. The precipitates are delivered, after drying, to optical emission spectrometry, X-ray fluorescence spectrometry [319–322] or activation analysis, or dissolved in small amounts of diluted mineral acids or organic solvents for trace determinations by various techniques. The carrier elements and organic matter are sometimes removed by liquid-liquid extraction, dry or wet oxidation, etc. prior to trace determinations.

The following should be considered in the *selection of a collector precipitate:*

(1) *Trace recoveries.* Several rules in coprecipitation phenomena described in Chap. 7.1.1 may be helpful in the selection of a carrier element and a precipitant, which give sufficiently high trace recoveries. For example, flocculent amorphous precipitates having large specific surface areas may be favorable for adsorption and occlusion of trace elements. Isomorphous mixed crystal formation works effectively in other cases.

(2) *Selectivity.* Masking frequently improves selectivity. For example, traces of Ag, Au and Hg are simultaneously collected on *p*-dimethylaminobenzylidenerhodanine precipitates, but Ag is selectively precipitated in the presence of cyanide or iodide ions [323]. In the carrier precipitation of trace impurities with hydroxide precipitates in ferrous and nonferrous metals, EDTA is effectively used to mask the precipitation of the matrix elements [324].

(3) *Ease of physical separation* of the precipitate from the mother liquor. Aging of the precipitate, which generally facilitates filtration, often reduces trace recoveries. When both filtration and centrifugation are difficult or impossible, flotation described in Chap. 10.2 can be employed.

(4) *Interferences* of the collector precipitates in the later separation or determination steps.

Typical collector precipitates are tabulated in Table 27.

7.2.2 Application of Carrier Precipitation

Carrier precipitation is extensively used in the enrichment of trace elements in fresh, sea and waste waters. Table 28 lists some examples. Enrichment factors of 10^3 are

Table 27. Typical collector precipitates

Collector precipitates	Elements collected
Hydroxides or hydrated oxides	
$Fe(OH)_3$	Ag, Al, As, Ba, Be, Bi, Cd, Ce, Co, Cr, Cu, Dy, Er, Eu, Ga, Gd, Ge, Ho, In, Ir, La, Lu, Mg, Mn, Mo, Nb, Nd, Ni, Np, Pa, Pb, Pd, Pr, Pt, Pu, Rh, Ru, Sb, Sc, Se, Sm, Sn, Sr, Ta, Tb, Tc, Te, Th, Ti, Tl, Tm, U, V, W, Y, Yb, Zn, Zr
$Al(OH)_3$	Be, Bi, Cd, Ce, Co, Cr, Cu, Dy, Er, Eu, Fe, Ga, Gd, Ge, Hf, Ho, Ir, La, Lu, Mg, Mn, Mo, Nb, Nd, Ni, P, Pb, Pm, Pr, Pt, Rh, Ru, Sc, Se, Sm, Sn, Tb, Th, Ti, Tl, Tm, U, V, W, Y, Yb, Zn, Zr
MnO_2	Al, As, Au, Bi, Cr, Cu, Fe, Ga, In, Mo, Nb, Pa, Pb, Po, Sb, Se, Sn, Te, Th, Tl, V
$Bi(OH)_3$	Cd, Co, Cu, Fe, Mn, Pb, Zn
$Sn(OH)_4$	Cd, Co, Cu, Fe, Zn
Sulfides	
CuS	Ag, As, Au, Bi, Cd, Fe, Ga, Hf, Hg, In, Mo, Nb, Pb, Pd, Po, Pt, Rh, Ru, Sb, Sn, Ta, Tc, Te, Ti, Tl, V, W, Zn, Zr
HgS	Ag, Au, Bi, Cd, Ga, Ge, In, Pb, Tl, Zn
CdS	Cu, Fe, Hg, In, Sb, Sn, Tl
PbS	Au, Cu, Pd, Pt, Tl
Other inorganic precipitates	
Te	Ag, Au, Bi, Hg, Mo, Pb, Pd, Po, Pt, Sb, Se, Sn, Ti
As	Ag, Bi, Hg, Mo, Pb, Re, Sb, Se, Sn, Te
Se	Au, Pd, Pt, Te
CaF_2	Al, Ce, Fe, Gd, Pb, Pu, Th, Y, Zr
YF_3	Ce, Dy, Er, Eu, Gd, Ho, La, Nd, Pm, Pr, Sm, Tb, Tm, Yb
LaF_3	Am, Ce, Np, Pu, Th, U
Organic precipitates	
Thionalide	Ag, As, Au, Hg, In, Os, Ta, Zn
Dithizone	Ag, Au, Cu, Hg
p-Dimethylaminobenzyl-idenerhodanine	Ag, Au, Pd
1-Nitroso-2-naphthol	Ce, Co, Fe, U, Zr
2-Mercaptobenzimidazole	Ag, Au, Hg, Sn, Ta
2-Mercaptobenzothiazole	Ag, Au, Hg
Copper oxinate	Al, Ca, Cd, Cu, Fe, Hg, Mg, Mn, Zn
Copper cupferrate	Bi, Fe, Hf, Mo, Nb, Sn, Ta, Ti, V, W, Zr

easily attained with greater than 90% yields for numerous heavy metals at the $\mu g/l$ level or lower. Most of alkali and alkaline earth elements remain in the solutions.

This technique is also useful in concentrating trace impurity elements at the ng/g or low $\mu g/g$ level in high-purity metals and other inorganic solid samples. Table 29 lists some examples. By proper selection of collector precipitates and masking agents, enrichment factors of greater than 10^3 are frequently achieved, which can be much improved by the use of the reprecipitation technique.

Carrier precipitation is widely used in radiochemical separations with isotopic or non-isotopic carriers.

Table 28. Carrier precipitation of trace elements in water samples

Collector precipitates or precipitants	Trace elements enriched	Determination techniques	Lit.
Inorganic			
$Fe(OH)_3$	As, Cd, Co, Cr, Cu, Ni, Pb, Se, V, Zn	Phot., AAS, OES, XRF, Gutzeit test	[325–332]
$Fe(OH)_3 + Ti(OH)_4$	V	AAS	[329, 333]
$Al(OH)_3$	Zr	Fluor.	[334]
$Zr(OH)_4$	As, Cr, Pb, Sb	AAS, Phot.	[335, 336]
$Mg(OH)_2$	Fe, Mn	AAS, Phot.	[337, 338]
MnO_2	Mo, Sb	Phot.	[339, 340]
CuS	Cd	Phot.	[341]
$SrCO_3$	Cd	AAS	[342]
$AlPO_4$	U	Phot.	[343]
CaF_2	U	Fluor.	[344]
Organic			
$Ca(COO)_2$	Sc	Phot.	[345]
2-Mercapto-benzimidazole	Au	Phot.	[346]
1-Nitroso-2-naphthol	U	Fluor.	[347]
α-Benzoinoxime	Mo	Phot.	[348]
Thionalide	Ag, As, Cu, Sb	Phot., NAA	[349–351]
α-Benzildioxime	Ni	Reflectance measurement	[352]
Oxine	Cu, Fe, Mn	NAA, Phot.	[353, 354]
5,7-Dibromo-8-hydroxyquinoline	Co, Cr, Cu, Fe, Mn, Zn	AAS, Phot.	[355]
Oxine + tannic acid + thionalide (In carrier)	Al, Be, Bi, Cd, Co, Cr, Cu, Fe, Ga, Ge, Mn, Mo, Ni, Pb, Ti, V, Zn	OES	[356]
PAN	Cr, Cu, Eu, Fe, Hg, Mn, Ni, Zn	XRF	[357, 358]
Polyvinylpyrrolidone + thionalide	Cd, Cu, Fe, Hg, Pb, Se, Sn, Te, Zn	XRF	[359]
6-Anilino-1,3,5-triazine-2,4-dithiol	Cd, Cu, Pb	XRF	[360]
DDTC	Cd, Cr, Cu, Fe, Mn, Pb, Sb, Ti, Zn	XRF, Phot.	[358, 361, 362]
Diethylammonium diethyldithio-carbamate	Cd, Cr, Cu, Hg, Ni, Pb	XRF	[363, 364]
APDC	As, Cd, Co, Cu, Fe, Ni, Pb, Zn	AAS, XRF	[330, 362, 365–367]
Iron dibenzyl-dithiocarbamate	U	XRF	[368]
TTA + naphthalene	Fe	Phot.	[369]

Table 29. Carrier precipitation of trace elements in high-purity metals and other inorganic solid samples

Matrices	Trace elements	Collector precipitate or precipitant	Determination techniques	Lit.
Cu	As	$Fe(OH)_3$	Phot.	[370]
Ag	Bi, Pb, Te	$Fe(OH)_3$	Polar.	[371]
Al	Mn	$Fe(OH)_3$	Phot.	[372]
Ni	Cu	$Fe(OH)_2$	Polar.	[373]
Cr	P	$Al(OH)_3$	Phot.	[374]
Ag	Bi	$Al(OH)_3$	Polar.	[375]
Fe	Sb, Ti	$Cr(OH)_3$	Polar.	[376]
Ag, Cd, Cu, Zn	Fe	$Cr(OH)_3$ + $Ti(OH)_4$	XRF	[377]
Na	Co, Cr, Fe, Mn, Ni	$La(OH)_3$	OES	[378]
Cu	As, Bi, Fe, Pb, Sb, Se, Sn, Te	$La(OH)_3$	AAS	[379]
Fe	Cr, Sn	$Be(OH)_2$	Polar.	[380, 381]
Al	Fe, Mn, Ti, Zn	$Zr(OH)_4$	AAS, Polar.	[382, 383]
Mg	Co, Cu, Fe, Zn	$Sn(OH)_4$	OES	[384]
Mo, W	Ti, Zr	$Co(OH)_2$	XRF	[385]
Al	Cr, Cu, Fe, Mg, Mn, Zn	$Ni(OH)_2$	AAS	[386]
Fe	Sb	MnO_2	Phot.	[387]
Cu	Sn	MnO_2	Polar.	[388]
Pb	Sb, Tl	MnO_2	Phot.	[389]
Ni	Bi, Pb	MnO_2	AAS	[390]
In	Au, Bi, Cd, Hg, Mo, Pd, Sb	CuS	OES	[239]
Ag	Au	Ag_2S	NAA	[391]
Uranium ores	Th	LaF_3	OES	[392]
Be, Ti, U, Zr	Rare earths	CaF_2 + MgF_2 + YF_3	OES	[393]
U, Zr	Rare earths	$ThF_4 \cdot NH_4F$	X-ray luminescence	[394]
U	Ag	TlI	Phot.	[395]
Ag, Cr, Cu, Mg, Ni, Zn	Pb	$BaCrO_4$	Polar.	[396]
Pb	Se	$PbSO_4$	Phot.	[397]
Ag, Cu, Hg, Ni	Pd	AgCN	OES, Phot.	[398, 399]
Cu, Pb, Heat-resisting alloys	Se, Te	As	Phot., AAS, XRF	[400–402]
Cu	Au	Te	Phot.	[403]
Telluric acid	Pb	Te	Phot.	[404]
Pb	Ag, Au, Bi, Cu, Pd	Pb	AAS	[405]
In	Bi, Fe, Hf, Mo, Nb, Sn, Ta, Ti, V, W, Zr	Cu cupferrate	OES	[239]
Steels	Zr	Fe cupferrate	Phot.	[406]
Gypsum	Al, Fe, Ti	Cupferron	XRF	[407]
Al	Bi, Cd, Co, Fe, In, Ni, Pb, Tl, Zn	APDC (Cu carrier)	AAS	[408]

Table 29 (continued)

Matrices	Trace elements	Collector precipitate or precipitant	Determination techniques	Lit.
K and Na salts	Cu, Fe, Mn, Ni, Pb, Sn, Zn	Oxine + thionalide	OES	[409]
KCl	39 elements	Oxine + tannic acid + thionalide (In carrier)	OES	[410]
Cd	Ag, Bi, Cr, Cu, Fe, In, Ni, Pb	α-Benzildioxime + CdS + MnO_2	AAS	[411]

8 Electrochemical Deposition and Dissolution

8.1 Electrodeposition on Solid Electrodes

By electrolysis, various elements in a solution can be deposited on a solid working electrode under proper conditions: e.g. Ag, Au, Bi, Cd, Co, Cu, Fe, Hg, Ni, Pb, Pd, Sb, Sn, Te and Zn on a platinum cathode as metals; Co, Mn, Ni, Pb and Tl on a platinum anode as oxides; and Cl, Br, I or S on a silver anode as halides or sulfide.

Materials used for working electrodes include platinum, platinum-iridium alloys, silver, copper, tungsten and carbon (graphite, pyrolytic graphite, glassy carbon, etc.). As materials for counter electrodes, platinum or platinum-iridium alloys are most common. It must be kept in mind that the platinum anode dissolves slightly in acidic or ammoniacal electrolytes and redeposits on the cathode. Carbon, silver (for electrolytes containing chloride ions) and lead (for ammoniacal electrolytes) are sometimes employed as counter anodes. The shapes of working and counter electrodes are straight or spiral wires, wire gauzes, sheets, rods, tubes, cups and crucibles. The material, shape, and physical isolation (diaphragm cells) of the counter electrode should be carefully considered to minimize undesired reactions at this electrode. Controlled potential electrolysis requires a *reference electrode* such as the saturated calomel, the silver chloride or the mercury(I) sulfate electrode. A reference electrode sometimes serves as a counter electrode, too. Fig. 22 shows two types of electrolysis cells. Besides con-

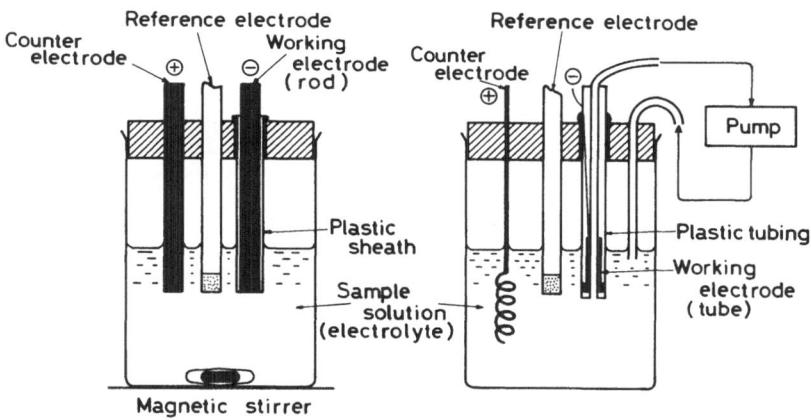

Fig. 22. Electrolysis cells

ventional static electrolysis in a glass or plastic beaker cell (Fig. 22 A), flow electrolysis is employed for the deposition of trace elements on the inner wall of a small tube [412, 413] (Fig. 22 B) or cup [414] electrode.

Electrolytes are prepared by adding the following reagents to the sample solution, if necessary [415].

(1) *Depolarizers*, which prevent undesired electrode reactions by being preferentially reduced at a cathode or oxidized at an anode, and maintaining the cathode potential less negative or the anode potential less positive than the potential at which the undesired reactions occur. For example, dissolution of a platinum or platinum-iridium anode in an electrolyte containing chloride ions can be prevented by addition of anodic depolarizers such as hydrazine and hydroxylamine. Anodic depolarizers are also effective in the cathodic deposition of cobalt to prevent its deposition as oxide on a platinum anode. To the contrary, cathodic depolarizers (HNO_3 and Cu^{2+}) are used in the anodic deposition of lead to prevent its deposition as metal on a platinum cathode.

(2) *Complexing agents*, which facilitate the electrolytic separation by shifting the deposition potential, or improve adhesion of the deposit to the electrode and smoothness of the surface of the deposit.

(3) *Buffers*, which minimize the pH change of the electrolyte during the electrolysis.

(4) *Surfactants*, which give smooth metal deposits.

Controlled potential electrolysis, where the potential of the working electrode is externally kept constant during the electrolysis with respect to a reference electrode, enables various elements having different deposition potentials to be separated from each other. When the potential of the working electrode exceeds the deposition potential by about 0.2 V, deposition yields of greater than 99.9% are generally obtained. The constant current or the constant applied voltage electrolysis is also useful in many applications. Even in these cases, the change in potential of the working electrode is limited during the electrolysis under proper conditions, e.g. by hydrogen evolution in acidic electrolytes.

Theoretically, in controlled potential electrolysis, the quantity Q of an element deposited on a working electrode from a stirred electrolyte increases with electrolysis time t according to Eq. (35).

$$Q = Q_0 \left(1 - e^{-kt}\right) \tag{35}$$

where Q_0 is the initial quantity of the element in the electrolyte and k is a constant, which is directly proportional to the surface area of the working electrode, inversely proportional to the volume of the electrolyte, and increased with agitation of the solution.

It is not difficult to deposit nanogram quantities of trace elements on a small solid electrode (surface area about 0.1 cm^2) from 10 ml of electrolyte in greater than 95% yields within reasonable time. Deposition of a definite portion of an element often suffices, e.g. in stripping voltammetry described later. Electrodeposition of trace elements on a solid electrode can be applied to extremely dilute solutions such as carrier-free solutions of radioactive nuclides. It must be kept in mind, however, that anomalous behavior is often observed [5]. For example, there are often considerable differences between deposition potentials observed and those calculated by the Nernst equation and standard potentials. In addition, great care must be taken to prevent losses due to adsorption of trace elements on surfaces other than the working electrode and redissolution of the deposit by rinsing. Contamination originating from counter and reference electrodes is another problem.

The trace elements deposited on a small solid electrode are determined *in situ* by various methods. In anodic or cathodic stripping voltammetry, trace elements are first deposited on an electrode at a controlled potential and then anodically or cathodically redissolved into the original (or another) electrolyte by scanning the electrode potential linearly with time by a polarograph, a potential-current curve (Fig. 23) being recorded. When a definite portion of an element is electrodeposited, the area or height of the peak is used for the determination. This highly sensitive technique is useful in the determination of impurities at the ng/g level or lower in natural waters and high-purity materials. For further details, see monographs [416–418].

Other *in situ* determination methods include electrothermal atomic absorption spectrometry, emission spectrography, mass spectrometry, X-ray fluorescence spectrometry, electron and ion microprobe techniques, and radioactivity measurements. High-purity carbon electrodes are ideal supports for neutron irradiation in a nuclear reactor for activation analysis. Some examples are shown in Table 30.

The trace elements deposited on an electrode can also be determined after dissolution in mineral acids, etc. [430].

Removal of over 99.9% of the matrix elements by electrodeposition on a large solid electrode may sometimes be useful.

8.2 Electrodeposition on Mercury Cathodes

Electrolysis with a mercury cathode [431, 432] has wide applicability in trace analysis. A large number of elements can be deposited even from acidic aqueous solutions because of the high overvoltage of hydrogen on mercury.

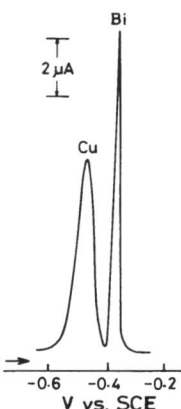

Fig. 23. Anodic stripping curve

Table 30. Electrodeposition of trace elements for *in situ* determinations (excluding stripping voltammetry)

Working electrodes	Deposited trace elements	Separated from	Determination techniques	Lit.
Carbon				
Rod	Ag	Zn	OES	[419]
	Bi, Cd, Cu, Hg, In, Pb, Tl	K, Mg	AAS	[420]
	Au	Cd	NAA	[421]
	Co, Cr, Cu, Hg, Ni, Zn	–	XRF	[422, 423]
Tube	Bi, Co, Fe, Zn	–	AAS, OES, NAA	[412]
	Co, Cr, Ni, Pb	Sea water	AAS	[413]
Crucible	Pb	Sea water	AAS	[414]
Platinum				
Wire	Ag	Cd	NAA	[424]
	Cd	Urine	AAS	[425]
Disk	Cu	–	EPMA	[426]
Tungsten				
Wire	Cd	Sea water	AAS	[427]
Copper				
Wire	Hg	Urine	AAS	[428]
Silver				
Disk	S	–	IMA	[429]

Various electrolysis cells have been reported for mercury cathode electrolysis, including those for micro or ultramicro sample solutions [433]. A mercury pool, or sometimes a solid electrode covered with mercury, is used as the cathode, and platinum, platinum-iridium alloys, or sometimes carbon, silver and lead, are used as the anode in the shape of straight or spiral wire, wire gauze or sheet. The material and shape of the counter electrode should be carefully selected to minimize undesired reactions at this electrode. Fig. 24 shows some types of electrolysis cells useful in inorganic trace analysis.

As electrolytes, dilute sulfuric or perchloric acid solutions are most frequently used. In general, the lower the acidity, the faster the deposition. Various electrolytes containing other mineral and organic acids can also be employed. Anodic depolarizers and pH buffers are sometimes added to the electrolytes.

The behavior of elements in mercury cathode electrolysis is summarized in Table 31.

8.2.1 Deposition of Trace Elements

Most commonly, the desired trace elements in dilute sulfuric or perchloric acid solutions are deposited in the mercury cathode by the constant current or the constant applied voltage electrolysis in the cell shown in Fig. 24 A. Typical operating condi-

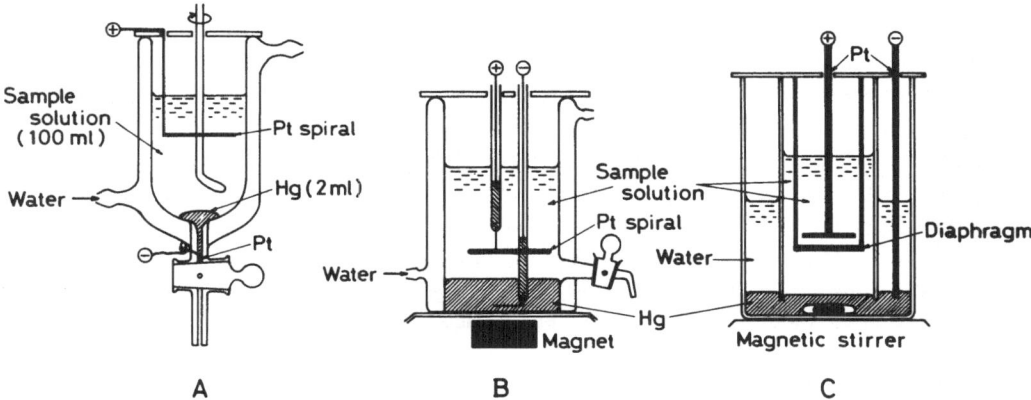

Fig. 24. Mercury cathode electrolysis cells. A: Cell for deposition of trace elements, B: Cell for removal of matrix elements, C: Double cell

Table 31. Behavior of elements in mercury cathode electrolysis

Using the electrolysis cells shown in Fig. 24 A, B	
(ca. 0.15 M sulfuric acid solution)	
Elements quantitatively deposited in the cathode	Ag, Au, Bi, Cd, Co, Cr, Cu, Fe, Ga, Ge, Hg, In, Ir, Mo, Ni, Pd, Po, Pt, Re, Rh, Sn, Tc, Tl, Zn
Elements quantitatively separated from the electrolyte but not quantitatively deposited in the cathode	As, Os, Pb, Se, Te
Elements incompletely separated	La, Mn, Nd, Sb, Ru
Elements remaining in the electrolyte	Alkali and alkaline earth elements, Al, B, Hf, Nb, P, Ta, Ti, U, V, W, Zr
Using the electrolysis cell shown in Fig. 24 C	
Elements which go to the outer compartment	Ba, Ca, Cs, K, Li, Na, Rb, Sr, Th, Ti, Zr
Elements remaining in the electrolyte	Be, Cr, La, Mg, U, V

tions are: total current of 1 to 2 A, current density at the cathode of 0.1 to 1 A/cm^2, and applied voltage of 7 to 20 V. The cathode potential in kept nearly constant in the presence of so-called oxidation-reduction (redox) buffers such as large amounts of U(IV)-U(III), Ti(IV)-Ti(III), V(III)-V(II) and H(I)-H(O) couples in the electrolyte. Simultaneous evolution of hydrogen on the cathode accelerates the deposition [434]. The deposition proceeds according to Eq. (35), and microgram quantities of elements are separated in greater than 95% yields within a few hours.

After completion of the electrolysis, the resulting dilute amalgam is taken out of the cell through the stopcock, and the desired trace elements are separated from the mercury by one of the following two methods. One is the evaporation of mercury at about 350 °C in a vitreous-silica boat in a stream of nitrogen. Microgram or nanogram quantities of metals such as Ag, Au, Co and Fe in a few milliliters of dilute amalgam can completely recovered in the residue. Some danger of loss exists for metals with a low melting point such as cadmium. The evaporation residue can be dissolved in a very small amount of mineral acids. The other technique [435] is anodic stripping, in which the desired trace elements are anodically dissolved in 5 to 10 ml of 0.1 to 0.5 N potassium sulfate or potassium chloride by controlling the potential of the amalgam 0.1 to 0.5 V lower than the dissolution potential of mercury. Recoveries greater than 95% are obtained for 10 to 1000 μg of Cd, Cu, Pb and Zn by this technique. Iron, Co and Ni remain in the amalgam. Since a dilute amalgam of less than 10^{-6} M has the same potential as that of pure mercury [436], it is estimated that about 0.06 μg of Zn, 0.1 μg of Cd, 0.2 μg of Pb or 0.06 μg of Cu remains in 1 ml of mercury after anodic stripping. It is very difficult to selectively dissolve the desired trace elements completely in acids or other appropriate solutions without introduction of appreciable amounts of mercury into the solutions.

The enrichment technique has been successfully applied to the spectrophotometric and polarographic determinations of impurities at the low μg/g level in various samples: Bi, Cd, Co, Cu, Fe, Ni, Pb and Zn in uranium and its compounds [437–439]; Cd, Co, Cu, Fe, Ni, Pb and Zn in vanadium compounds [435]; Co in titanium and zirconium [440]; and Cd, Co, Cu, Fe and Ni in high-purity magnesium [441] and aluminum [442]. Generally, greater than 95% yields are attained with enrichment factors of over 10^4. It must be kept in mind that microgram quantities of platinum dissolve from the anode in dilute sulfuric acid electrolyte, redeposit on the cathode together with the desired trace elements, and may interfere with later spectrophotometric or polarographic determinations [437, 443].

Mercury cathode electrolysis with the double cell shown in Fig. 24 C is applied to the enrichment of Na at the μg/g level in magnesium prior to neutron irradiation for activation analysis [444]. Electrolysis for 4 h in a hydrochloric acid-citric acid electrolyte gives yields of about 95% and enrichment factors of about 10^3.

Controlled potential electrolysis with a mercury cathode is also useful. For the separation of low μg/g of Pb and Cd in zinc-base alloys [445], Pb, Cd and Cu are deposited in a mercury cathode controlled at -0.9 V vs. SCE from hydrochloric acid solution containing hydrazine as anodic depolarizer. Lead and Cd are then anodically dissolved from the resulting amalgam at -0.35 V vs. SCE in 0.1 M potassium chloride solution containing hydrazine, leaving Cu in the amalgam, and determined by polarography. Such impurities as Co, Cu, Fe, Ni, Pb and Zn in high-purity niobium are enriched by controlled potential mercury cathode electrolysis in a hydrofluoric acid solution in a Teflon cell [446]. After the electrolysis, the mercury is evaporated off under vacuum in a small graphite cup, which is then used in electrothermal atomic absorption spectrometry.

Traces of Cd, Pb and Zn in sea water and reagent-grade potassium chloride are partly deposited on a hanging mercury drop electrode, and after removal of mercury by evaporation, determined by electrothermal atomic absorption spectrometry [447]. Similarly, Pb and Cd in sea water are deposited on a mercury-coated graphite tube electrode for the electrothermal atomic absorption spectrometric determination [448]. Another example is the application to radiochemical separations for neutron activation analysis of sea water [449]. Finally, stripping voltammetry with a hanging mercury drop electrode, tiny mercury pool electrode, or mercury-plated solid electrode offers many highly sensitive determination methods [416–418].

8.2.2 Deposition of Matrix Elements

The electrolysis cell shown in Fig. 24 B is conveniently used for removing matrix elements from a sample solution by deposition on a mercury cathode with an electrolysis current of 10 to 20 A. The magnetic field stirs the mercury-electrolyte interface vigorously in combination with the electric current, removes deposited ferromagnetic metals continuously from the mercury surface, and thus accelerates the rate of the deposition of heavy metals [450]. There are a number of other types of cells without magnetic field; other means of stirring are often provided. After completion of the electrolysis, the solution containing the desired trace elements is removed through the stopcock and used for the determination.

This technique is useful in separating trace elements such as Al, B, Ca, Mg, Ti, V, W and rare earths in iron, steel, nickel and other metals and alloys. For example, low μg/g of rare earths in stainless steel are separated from the matrix elements and determined by emission spectrography [451]. Similarly, Al in iron, steel and ferrous alloys is separated for atomic absorption spectrometry [452]. For the separation of B at the low μg/g level in nickel, the sample piece itself is used as anode in a mercury cathode electrolysis [453]. A water-cooled platinum crucible containing small amounts of mercury at the bottom is used as electrolysis cell. The dissolution of the sample and the removal of the matrix element take place simultaneously in 0.01 M sulfuric acid. There is no danger of contamination from the anode material. In addition, contamination due to impurities in sulfuric acid is further reduced than the conventional technique, where larger amounts of the acid are required for the dissolution of the metal. After the electrolysis, B is determined by spectrophotometry. A similar technique using 0.5 M perchloric acid as electrolyte is applied to the polarographic determination of Al down to 0.3 μg/g in iron [454].

Removal of the matrix element by controlled potential electrolysis is a useful technique prior to the polarographic determination of trace elements. For the enrichment of Ni and Zn at the low μg/g level in copper and its compounds [455], Cu is deposited in a mercury cathode controlled at -0.85 V vs. SCE from an aqueous ammonia-ammonium chloride buffer solution containing hydrazine in a nitrogen atmosphere. After removal of about 99.99% of Cu, Ni and Zn remaining in the electrolyte are determined by polarography. In case Ni exists in larger amounts compared with Zn, Ni is removed by electrolysis with a mercury cathode at -1.20 V vs. SCE prior to the determination of Zn. A similar technique is applied to the separation of low μg/g of Zn in cadmium [415].

8.3 Spontaneous Electrochemical Deposition

Spontaneous electrochemical deposition without an external source of electrical energy is sometimes useful.

Noble metals (Ag [456, 457], Au [458, 459], Pt [460] and Pd [461]) can be deposited on mercury globules (1 to 4 mm diam.) by magnetically stirring 2.5 to 500 ml of acidic or ammoniacal sample solutions with 0.5 to 2 ml of mercury in a beaker for a half to a few hours. Such base metals as Cu, Pb and Fe remain in the solution. The resulting dilute amalgam is then separated from the sample solution by decantation and heated to evaporate off the mercury. A more rapid collection (within 1 min) of silver is effected by the use of a mercury-in-water emulsion (1 to 4 μm particle diameter) under ultrasonic irradiation [462]. By these techniques, Ag and Au at the ng/g or low μg/g level in copper and lead metals are preconcentrated for the spectrophotometric or atomic absorption spectrometric determination, with recoveries of better than 95% and enrichment factors of 10^4 to 10^6. A similar technique is used in separation of Au in lead concentrates [463].

Mercury, 0.1 to 1.5 μg/l, in dilute (about 1 M) hydrochloric acid is collected as amalgam on a silver wire coil for 5 to 6 h and determined by atomic absorption spectrometry [464]. This technique is applied to fresh waters without interferences from commonly coexisting substances, except for iron (>5 mg/l) and sulfide which cause low recovery of mercury. Similarly, a micro-column of copper powder is used for the preconcentration of mercury [465].

8.4 Anodic Dissolution

A controlled potential electrolysis using a steel plate sample as anode enables the iron matrix to be dissolved, leaving oxides, carbides, nitrides, sulfides, phosphides and other inclusions in the steel as residue [146, 263]. Electrolytes useful for this purpose include 10% acetylacetone-1% tetramethylammonium chloride-methanol, 15% sodium citrate-30% citric acid-1.2% potassium bromide aqueous solution (pH = 3.0), and 7% hydrochloric acid-3% iron(III) chloride-ethylene glycol. After the electrolysis, the residue is separated by filtration, further separated by selective dissolution, magnetic separation and sieving, if necessary, and then observed by optical and electron microscopes or analyzed by various determination methods. Etching of the surface layer (about 1 μm) of a steel sample by anodic dissolution is useful for the *in situ* observation and analysis of inclusions by various microscopes and microanalyzers [466].

9 Sorption, Ion Exchange and Liquid Chromatography

The enrichment techniques discussed in this chapter are based on the distribution of substances between a solution and a solid sorbent or ion exchanger by mechanisms such as physical sorption, ion exchange, complex formation and other chemical reactions, on, or in, the sorbents. Liquid extractants supported on a suitable inert solid are also included here, because they behave similarly as solid sorbents by a mechanism of liquid-liquid partition in liquid chromatography. Generally, the degree of sorption at equilibrium is described by the distribution coefficient, i.e. the total amount of an element in unit quantity (g or ml) of sorbent divided by the total amount of the same element in unit quantity (ml) of solution. The weight distribution coefficient, D (ml solution/g dry sorbent), is related to the volume distribution coefficient, D_v (ml solution/ml sorbent bed), by the relationship $D = D_v/\rho$, where ρ is the sorbent bed density.

9.1 General Procedures

9.1.1 Batch Operation

In this operation, sorption of the desired trace elements is carried out simply by immersing a sorbent in a sample solution in a suitable vessel. To attain the sorption equilibrium as quickly as possible, the solution is agitated mechanically or ultrasonically. The sorbent is then separated from the solution by decantation or filtration. If necessary, the sorbent is washed with a solution of appropriate composition, which removes undesired elements without desorption of the desired trace elements. Batch operation is useful, especially when the distribution coefficients of the desired trace elements are very large or column operation discussed in Chap. 9.1.3 is difficult. Desorption is carried out similarly as sorption using an appropriate solution. Without desorption, the sorbent containing the desired trace elements can be directly employed as samples for X-ray fluorescence spectrometry and neutron activation analysis, or incinerated or dissolved in acids or other solvents before the determination.

9.1.2 Filtration through a Permeable Sorbent Disk

A sample solution is filtered through a permeable sorbent disk such as a sheet of ion exchange paper and a thin layer of fine sorbent particles to selectively collect the desir-

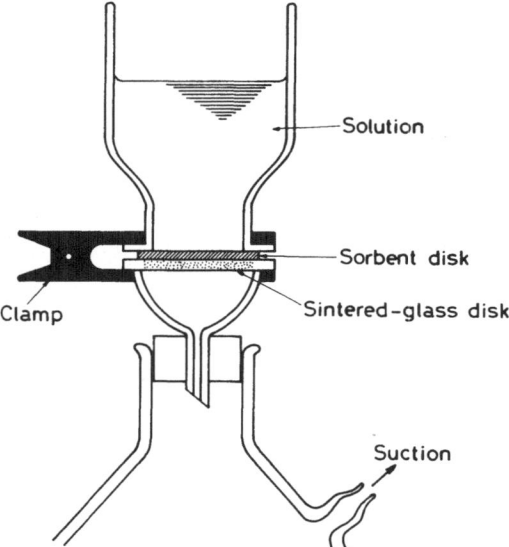

Fig. 25. Filtration apparatus

ed trace elements. A typical filtration apparatus is shown in Fig. 25. To achieve the maximum trace recoveries, sufficiently slow flow-rates should be used, or the filtration must be repeated until the sorption equilibrium is established. The operation is simple, but relatively large distribution coefficients and sorption rates are generally required. Washing and desorption are carried out likewise with suitable solutions. Without desorption, the disk can be used for the trace determination, directly or after decomposition, as described in Chap. 9.1.1.

9.1.3 Column Operation and Chromatography

This operation is most frequently used. A sample solution is passed through a column packed with a sorbent (Fig. 26) to collect either the matrix or the desired trace elements or sometimes both. In the first case, the percolated solution, combined with washings, is used for the determination of the desired trace elements. In the second and the third cases, after washing of the column, the trace elements are eluted from the column with eluents, chromatographically as shown in Fig. 27.

The plate theory [467] explains chromatographic elution curves and is very useful in designing chromatographic separations. In this theory, it is assumed that:

(1) The column consists of a great number of "theoretical plates" (Fig. 28), each containing V_s (ml) of the sorbent (stationary phase) and V_m (ml) of the eluent (moving or mobile phase), which are the same in each plate and remain constant during the elution step.

(2) In each plate, the following equilibrium is established instantaneously between the two phases:

$$C_{s,p} = D_v C_{m,p} \tag{36}$$

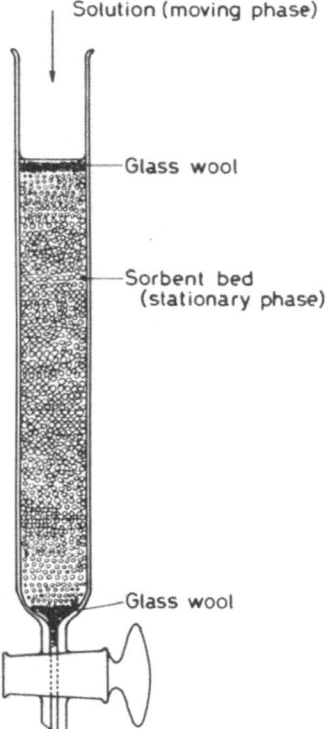

Fig. 26. Chromatographic column

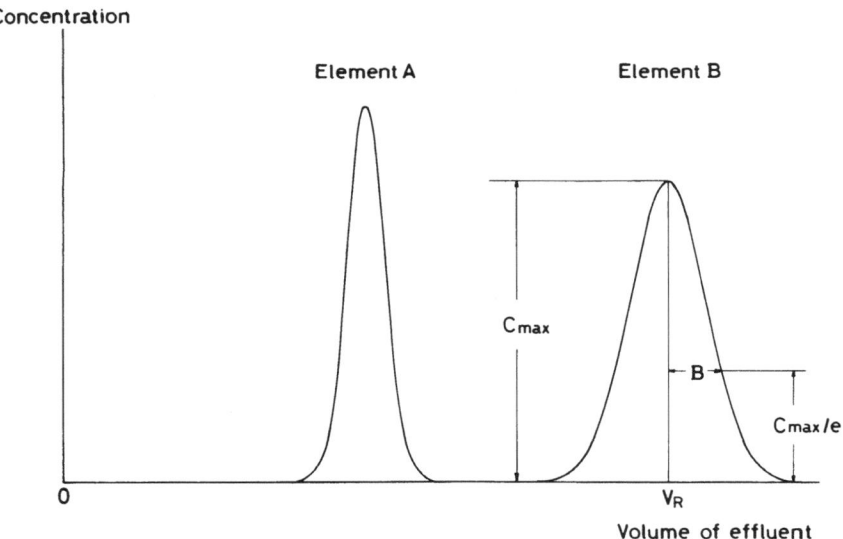

Fig. 27. Chromatographic elution curve

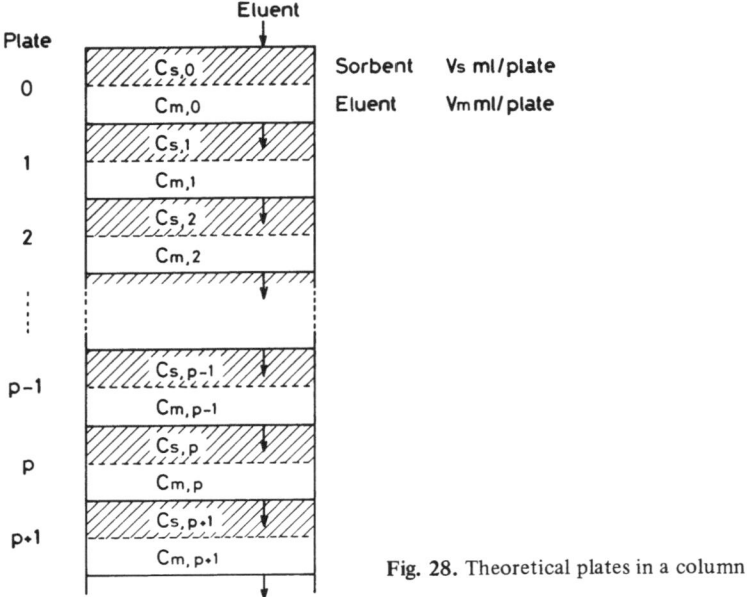

Fig. 28. Theoretical plates in a column

where $C_{s,p}$ and $C_{m,p}$ are concentrations of an element in the stationary and the moving phases, respectively, in plate p, and D_v is a volume distribution coefficient independent of the concentrations.

When dV (ml) of the eluent passes through the column, the conservation of mass in plate p requires:

$$(C_{m,p-1} - C_{m,p})\, dV = V_m\, dC_{m,p} + V_s\, dC_{s,p} \tag{37}$$

Elimination of $C_{s,p}$ using Eq. (36) gives:

$$\frac{dC_{m,p}}{dV} = \frac{C_{m,p-1} - C_{m,p}}{V_m + D_v V_s} \tag{38}$$

If the element is initially present only in plate 0, the solution of Eq. (38) is

$$\frac{C_{m,p}}{C_0} = \frac{v^p\, e^{-v}}{p!} \tag{39}$$

where

$$v = V/(V_m + D_v V_s) \tag{40}$$

and $C_{m,0} = C_0$ at $v = 0$. Equation (39), i.e. the Poisson distribution, is approximated by the Gaussian normal distribution for sufficiently large values of p and $|v - p| \ll p$:

$$\frac{C_{m,p}}{C_0} = \frac{1}{\sqrt{2\pi p}} \ e^{-(v-p)^2/2p} \tag{41}$$

Eq. (41) represents the equation of the elution curve (p = const.) as well as the distribution in the column of the element (v = const.). Thus the retention volume V_R, half the band width B, and maximum concentration C_{max} in Fig. 27 are obtained as follows:

$$V_R = I + D_v S \tag{42}$$

$$B = \sqrt{2/N} \ V_R \tag{43}$$

$$C_{max} = \sqrt{N/2\pi} \ Q/V_R \tag{44}$$

where I is the interstitial volume in the column (ml), S is the total volume of sorbent in the column (ml), N is the number of theoretical plates in the column, and Q is the total quantity of the element. According to Eq. (43), large numbers of theoretical plates are required to minimize overlap of chromatographic peaks. In general, with a given column length, the number of theoretical plates increases with decreasing particle size of the sorbent, decreasing flow rate and viscosity of the eluent, and increasing temperature.

In actual elution curves, broad asymmetric peaks with diffuse trailing edges frequently appear owing to the presence of large amounts of sorbed elements and non-linear isotherms, i.e. nonconformity to Eq. (36).

In favorable cases, the elution curve can be recorded continuously by means of an *in-line optical or electrochemical detector* at the bottom of the column and the area of each peak is measured for the trace determination. Frequently, this technique is not applicable because of insufficient sensitivity of the detector as well as a large background. Therefore, generally, appropriate fractions of the effluent are collected manually or by a fraction collector, concentrated by evaporation, if necessary, and analyzed for the desired trace elements. Automation of column operation is rather easy [468].

Backward-flow elution is sometimes applied to minimize the eluent volume required for desorption. In other cases, without elution, the sorbent containing the desired trace elements as such is used for the determination, after drying or decomposition, as described in Chap. 9.1.1.

Chromatographies on plane surfaces instead of columns, i.e. paper and thin-layer chromatographies, are sometimes used as enrichment techniques in trace analysis of microsamples.

9.2 Separation with Ion Exchange Resins

Among sorbents, synthetic ion exchangers, especially ion exchange resins, have widest applicability in enrichment of trace elements. There are several monographs dealing with analytical application of ion exchangers [150, 469–471].

Table 32. Most frequently used ion exchange resins

Resin	Strong-acid cation exchange resin	Strong-base anion exchange resin	Chelating resin
Functional group	Sulfonic acid $-SO_3^-$	Quaternary ammonium $-CH_2N^+(CH_3)_3$ or $-CH_2N^+(CH_3)_2(C_2H_4OH)$	Iminodiacetate $-CH_2N\big\langle{}^{CH_2COO^-}_{CH_2COO^-}$
Exchangeable pH range	0–14	0–14	6–14
Approximate specific capacity meq/g dry resin meq/ml resin bed	5 2	3 1	0.5
Thermal stability	Up to 150 °C	Up to 50 °C (OH-form) and 150 °C (Cl- and other forms)	Up to 75 °C
Approximate order of selectivity	$Th^{4+} > Al^{3+} > Mg^{2+} > Na^+$; $Ac^{3+} > La^{3+} > Y^{3+} > Sc^{3+} > Al^{3+}$; $Ra^{2+} > Ba^{2+} > Pb^{2+} > Sr^{2+} > Ca^{2+} > Cd^{2+} \gtrsim Zn^{2+} > Cu^{2+} \gtrsim Ni^{2+} \gtrsim Co^{2+} \gtrsim Mg^{2+} \gtrsim UO_2^{2+} > Be^{2+}$; $Tl^+ > Ag^+ > Cs^+ > Rb^+ > NH_4^+ \simeq K^+ > Na^+ > H^+ > Li^+$	$I^- > HSO_4^- > ClO_3^- > NO_3^- > Br^- > CN^- > HSO_3^- > NO_2^- > Cl^- > OH^- > F^-$	$Cu^{2+} > Pb^{2+} > Fe^{3+} > Al^{3+} > Cr^{3+} > Ni^{2+} > Zn^{2+} > Ag^+ > Co^{2+} > Cd^{2+} > Fe^{2+} > Mn^{2+} > Ba^{2+} > Ca^{2+} > Na^+ > K^+$
Inorganic impurities (μg/g in purified resins)	Fe < 1, Cu < 0.8, Ni < 0.05, Pb < 0.2, Al < 15, Total ash < 500	Fe < 0.5, Cu < 0.2, Ni < 0.05, Pb < 0.005, Al < 5, Total ash < 600	
Trade names	Dowex 50 and 50W; AG 50 and 50W; Amberlite IR-120, CG-120; ZeoKarb 225; Diaion SK	Dowex 1, Dowex 2; AG 1, AG 2; Amberlite IRA-400, CG-400, IRA-410; Diaion SA	Dowex A-1; Chelex 100

9.2.1 Ion Exchange Resins

An ion exchange resin is an insoluble but permeable synthetic polymer containing ionizable functional groups. Table 32 tabulates three types of ion exchange resins most frequently used in inorganic trace analysis. All of these use crosslinked styrene divinylbenzene copolymer as resin base (Fig. 29). As the degree of crosslinkage, i.e. percentage of divinylbenzene, increases, the wet volume capacity and the resistance to shrinking and swelling increase and the equilibrium rate and the permeability to large molecules decrease. Highly crosslinked resins (e.g. 8% divinylbenzene) with relatively small pores (microreticular resins) in the shape of spherical beads of 50 to 400 mesh (297 to 37 μm) are most commonly used.

Careful washing (conditioning) of commercial resins (except for purified ones) before use, with acids, alkalies, complexing agents, water and organic solvents, is very important to remove inorganic and organic impurities resulting from the manufacturing process. Also, it must be kept in mind that resins are sometimes decomposed in contact with solutions and cause contamination problems; for example, decomposition products from strong-base anion exchange resins interfere with the polarographic determination of trace metals.

Besides conventional beads, ion exchange resins are available in the forms of macroreticular or macroporous resin beads, membranes, impregnants of paper and membrane filters (e.g. ion-exchange resin-loaded papers) or foamed plastics such as open-cell polyether-type polyurethane foams [472, 473], and thin surface layers on solid inert spherical cores (pellicular and superficially porous ion exchangers for high performance liquid chromatography) [474].

There are other types of ion exchange resins having different resin bases and ionizable functional groups from those described above, though their analytical applications are limited. Those commercially available include moderately strong-acid cation exchange resins (crosslinked polystyrene with $-PO(OH)_2$ groups), weak-acid cation exchange resins (polymerized acrylic acid with $-COOH$ groups), weak-base anion exchange resins (crosslinked polystyrene with $-CH_2NH(CH_3)_2OH$ and $-CH_2NH_2CH_3OH$ groups), and ion exchange resins having both acidic and basic groups (ion-retardation or snake-cage resin, linear polyacrylic acid with $-COOH$ groups, trapped in crosslinked polystyrene with $-CH_2N(CH_3)_3Cl$ groups). Chelating resins [475] with the following functional groups are also useful in inorganic trace analysis: arsonic acid [476, 477],

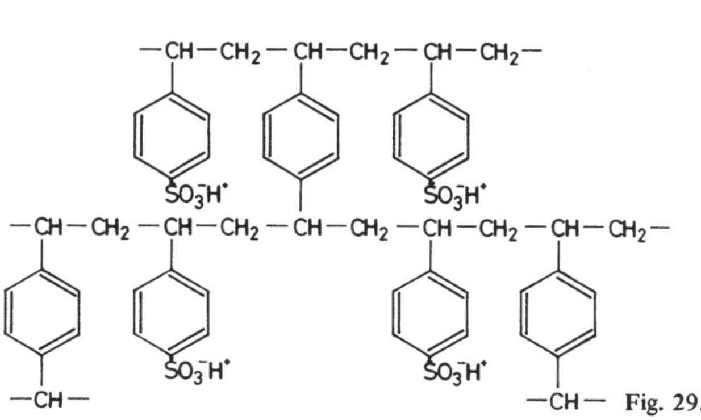

Fig. 29. Strong-acid cation exchange resin

oxine [478], dithiocarbamate [479–481], nitrosonaphthol [482, 483], amidoxime [484], etc. [485–487].

9.2.2 Ion Exchange Reactions and Equilibria

The ion exchange resin contains numerous fixed positive or negative charges and mobile counter ions of opposite charge, which maintain electrical neutrality in the resin and can be replaced reversibly by other ions of the same charge when the resin is immersed in a solution. For example, when a strong-acid cation exchange resin with mobile hydrogen ions (called an H-form cation exchanger) is immersed in an aqueous solution of sodium chloride, the following essentially reversible reaction takes place:

$$\text{Resin-SO}_3^-\text{H}^+ + \text{Na}^+ \text{(solution)} \rightleftharpoons \text{Resin-SO}_3^-\text{Na}^+ + \text{H}^+ \text{(solution)}$$

The equilibrium is expressed by

$$K_H^{Na} = \frac{[\text{Na}^+]_{\text{resin}} \, [\text{H}^+]_{\text{solution}}}{[\text{H}^+]_{\text{resin}} \, [\text{Na}^+]_{\text{solution}}} \tag{45}$$

The selectivity coefficient, K_H^{Na}, is a measure of affinity of the ion to the resin, although it is not constant because of great change in activity coefficients in the resin phase. If the sodium concentration is low and the ion exchange capacity of the resin is large enough, the distribution coefficient is independent of the sodium concentration. The same discussion is applied to cation exchange involving multivalent or complex ions as well as anion exchange.

The sorption behavior of an element on an ion exchange resin depends greatly on the chemical state of the element and the composition of the solution as well as the nature of the ion exchanger. Use of inorganic and organic complexing agents, highly concentrated solutions, and aqueous-organic mixed solvents often markedly improves the selectivity of ion exchange separations. Sometimes, besides ion exchange and chelation, other chemical reactions such as oxidation-reduction and precipitation in the resin phase play an important role in the sorption.

Formation of *metal complexes*, especially negatively charged complex ions, extends greatly the applicability of ion exchange separation of metal elements. For example,in hydrochloric acid solutions [488], various metals excluding alkali and alkaline earth metals, Ac, Al, Ni, Th, Y and rare earths form negatively charged chloride complexes and sorb on strong-base anion exchange resins. There is no simple relationship between the distribution coefficient and the hydrochloric acid concentration (0 to 12 M) for various elements. Replacement of hydrochloric acid with chlorides of Al, Ca, Li, Mg, etc., increases the distribution coefficients of the sorbed elements, especially at high chloride concentrations. Anomalous sorption of metals often occurs in concentrated solutions. For example, many metals including Bi, Ca, Fe, Ga, Hf, Mo, Sc, Sr, Th, Ti, U, W, Y, Zr and rare earths sorb strongly on strong-acid cation exchange resins from concentrated hydrochloric or perchloric acid solutions beyond expectation [489]. From dioxane-hydrochloric acid or dioxane-nitric acid mixtures, alkali and alkaline earth elements are selectively sorbed on an anion exchange resin at higher dioxane concentrations [490].

Distribution coefficients of various elements on strong-acid cation exchange resins in the following solutions have been extensively studied: hydrochloric acid [489, 491], hydrobromic acid [492], perchloric acid [489], nitric acid [493], sulfuric acid [493], hydrochloric acid-organic solvent media [494], hydrobromic acid-organic solvent media [495] and nitric acid-organic solvent media [496]. Distribution coefficients of various elements on strong-base anion exchange resins in the following solutions have been reported: hydrochloric acid [488], hydrofluoric acid [497], nitric acid [498], sulfuric acid [499], hydrochloric acid-hydrofluoric acid [500], nitric acid-hydrofluoric acid [501], acetic acid [502] and hydrobromic acid-organic solvent media [503]. Some of these data are given in Appendix A. 3.

Various metal ions are selectively sorbed on strong-base anion exchange resins from solutions containing chelating agents such as 8-quinolinol-5-sulfonic acid, or on anion exchange resins previously treated with chelating agents, which are prepared more easily than chelating resins with anchored chelating groups [504—514].

9.2.3 Sorption of Trace Elements on Ion Exchange Resins

The column operation is most frequently used. As shown in Fig. 30, trace elements in a sample solution of relatively large volume are quantitatively sorbed on a small column packed with ion exchange resin beads, provided that the distribution co-efficients of the trace elements are large enough. The sorbed trace elements are then eluted from the column with small amounts of eluents, and appropriate fractions, in which the trace elements are concentrated, are used for the trace determination. Chromatography using an in-line conductimetric detector is sometimes useful, where an additional resin column for background removal is placed between the separation column and the detector [515]. For example, in the chromatographic elution of mixtures of halides and other anions on an OH-form anion exchange resin column,

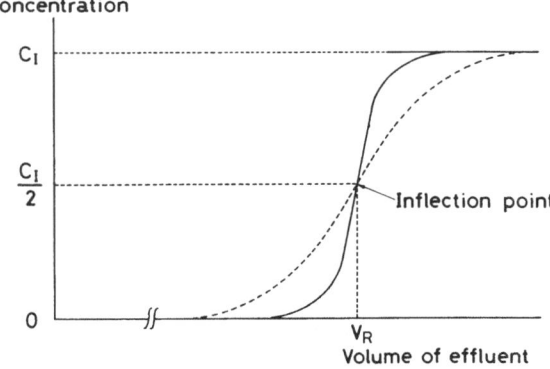

Fig. 30. Breakthrough curves. C_I: Concentration of element in influent. (————) Column of many theoretical plates. (————) Column of fewer theoretical plates

Table 33. Sorption of trace elements in water samples on ion exchange resins

Resin	Operation*	Trace elements	Sorbed from	Desorbed with	Determination techniques	Lit.
Cation exchange resins						
H-form	C	Cu	Water	5 M HCl	Phot.	[521]
H-form	C	Ag	Water	Boiling 1 M Na_2SO_3	Phot.	[522]
H-form	F	Co	Water	Hot HCl	AAS	[523]
H-form	B	Cr(VI)	River and sea waters + 1,5-diphenyl-carbohydrazide	–	Phot. (resin phase)	[518]
NH₄-form	C	Cs, Rb	Sea water	–	NAA	[524]
Cu(II)-form	C	$Fe(CN)_6^{3-}$, $Fe(CN)_6^{4-}$	Water	Aq. NH_3	AAS	[525]
PAN-loaded	B	Ni	River and sea waters	–	Phot. (resin phase)	[517]
Anion exchange resins						
Cl-form	C	Si	Water + HF	H_3BO_3	Phot.	[526]
Cl-form	C	NO_3^-	Water	1% NaCl	Phot.	[527]
Cl-form	C	Au, Bi, Cd	Sea water (0.1 M HCl)	0.25 M HNO_3, Water (Ashing)	OES, Phot., AAS	[528–530]
Cl-form	C	Ag	Sea water + NH_4 SCN	Hot 0.4 M thiourea	NAA	[531]
Cl-form	F	Hg	Water + HCl	–	NAA	[532]
Cl-form	B	Hg	Water + HNO_3 + NH_4 SCN (Reduction)		AAS	[533]
Cl-form	C	Cd, Cu, Mn, Pb, Zn	Water + 8-quinolinol-5-sulfonic acid	2 M HNO_3	AAS	[513]
Br-form	C	Cd, Cu, Pb	Water + HBr + ascorbic acid	1 M HNO_3	AAS	[534]
SCN-form	C	Co	Sea water + NH_4 SCN	2 M $HClO_4$	Phot.	[535]
SCN-form	C	Ti	Sea water + NH_4 SCN	2 M HCl-1.5% H_2O_2	Phot.	[536]
Molybdate-form	B	PO_4^{3-}, Silicic acid	Water	–	Phot. (resin phase)	[537]
Acetate-form	C	F	Water	0.1 M acetic acid –0.005 M Be	Phot.	[538]
Citrate-form	C	Th, U	Water + sodium citrate + ascorbic acid	HCl	Phot., Fluor.	[539]

Table 33 (continued)

Resin	Operation*	Trace elements	Sorbed from	Desorbed with	Determination techniques	Lit.
Zincon-loaded	C	Cu, Hg	River water	–	NAA	[507]
Batho-cuproine disulfonic acid-loaded	C	Hg	River water	–	NAA	[507]
Mixed cation and anion exchange resins						
Na- and Cl-forms	B	Cu	Water + 8-quinolinol-5-sulfonic acid	EDTA	AAS	[540]
Chelating resins						
NH_4-form Chelex 100	C (or B)	Cd, Co, Cr, Cu, Fe, Mn, Ni, Pb, U, Zn	Sea and waste waters	2.5 M HNO_3	AAS, OES, SSMS, NAA, Phot., Fluor.	[541–549]
NH_4-form Chelex 100	C	Cu, Mn, Zn, Rare earths	Sea and river waters	–	NAA	[550, 551]
Ca-form Chelex 100	C	Cd, Cu, Pb, Zn	River water	2 M HNO_3	ASV, AAS	[552]
Chelex 100 membrane	F	Co, Cu, Mn, Ni, Pb, Rb, Sr, Zn	Water	–	XRF	[553]
Dithio-carbamate resin	C	Hg	River and sea waters	Hot conc. HNO_3	AAS	[554]
Poly(acryl-amidoxime) resin	B	Cd, Cu, Fe, Pb, Zn	Sea and pond waters	6 M HCl	AAS	[484]

* B: Batch, C: Column, F: Filtration

the background due to eluents (NaOH, Na$_2$CO$_3$, NaHCO$_3$) is suppressed by neutralizing with an H-form cation exchange resin column.

NaX + Resin-H → Resin-Na + HX
NaOH + Resin-H → Resin-Na + H$_2$O

Likewise, in the cation exchange chromatographic elution of mixtures of alkali and alkaline earth elements, the neutralization of acids used for elution is effected by passing through an OH-form anion exchange column prior to the detector.

MCl + Resin-OH → Resin-Cl + MOH
HCl + Resin-OH → Resin-Cl + H$_2$O

Batch operation and filtration through a resin-loaded paper or membrane filter are also used.

Desorption of the desired trace elements is sometimes omitted. Thus, when the quantitative desorption of the desired trace elements, e.g. Au and Pt, is difficult with commonly available eluents, the resin is incinerated before the trace determination. The resin containing the desired trace elements can be irradiated in a nuclear reactor for neutron activation analysis. Pelletized resin beads, resin membranes, and resin-loaded paper or membrane filter disks are suitable supports for the direct determination of the sorbed trace elements by X-ray fluorescence spectrometry and proton-induced X-ray emission analysis [516]. Direct spectrophotometry of ion exchange resin beads containing the trace elements has been developed [517—520].

Table 33 lists examples of enrichment of trace elements in water samples by sorption on ion exchange resins.

Trace elements at the μg/l or low mg/l level in various inorganic solution samples are also enriched by this technique. Some examples follow. A few mg/l of Sn in hydrogen peroxide solutions is sorbed on an H-form cation exchange resin, eluted with 5 M hydrochloric acid, and determined by polarography [555]. Platinum in chlorate cell liquor down to 0.01 mg/l is sorbed on a Cl-form anion exchange resin from hydrochloric acid medium [556]. The resin is then incinerated and the Pt is determined by spectrophotometry. As little as 0.1 mg/l of U in barren uranium leach liquors (sulfate effluent) is sorbed on an SO$_4$-form anion exchange resin, and X-ray fluorescence spectrometry is carried out on the resin [557].

This technique is also applicable to organic or biological solution samples. Copper in mineral oils is sorbed on an H-form cation exchange resin from a one-to-one mixture of the sample and 2-propanol (sometimes plus a small amount of benzene to attain miscibility), eluted with 10% sulfuric acid followed by water, and determined by spectrophotometry down to 0.1 μg/g [558]. As little as 0.1 μg/g of Cu in milk is sorbed on an H-form cation exchange resin, followed by elution with 6% hydrochloric acid for polarography [559]. Ten trace metals in urine are sorbed on a poly(dithiocarbamate) resin column and then recovered by digestion with a one-to-one nitric-sulfuric acid mixture for inductively coupled plasma-optical emission spectrometry [480].

Table 34. Sorption of trace elements in high-purity metals and other inorganic solid samples on ion exchange resins

Matrices	Trace elements	Sorbed from	Desorbed with	Determination techniques	Lit.
Cation exchange resins					
Li	Ca	1–2 M LiCl	1 M HCl	Phot.	[560]
Zr	Rare earths	HF	6 M HCl	OES	[561]
Mo, W	Co, Cu, Fe, Ni, Zn	HF-H_2O_2	2 M HCl	Phot., ASV	[562]
Mo	Co, Cu, Fe, Mg, Mn, Ni, Pb, Zn	H_3PO_4-HNO_3-H_2O_2	6 M HCl	XRF, AAS	[563]
Ta, Nb	Cu, Mn	HF-HNO_3	6 M HCl	Phot., OES	[564, 565]
Fe, Zn	Al	Tetrahydrofuran-HCl	6 M HCl	Phot.	[566]
P	Cd	3 M H_3PO_4	4 M HCl	Polar.	[567]
Anion exchange resins					
Al, Mg	Cd, Co, Cu, Fe, Zn	HCl	HCl, HNO_3	Phot., Polar.	[442, 568]
Ni	Co	9 M HCl	0.1 M HCl or 1 M HNO_3	Phot.	[569]
Th	Cd	HCl	1 M HNO_3	Polar.	[217]
Cd, Cu, Zn	Au	HCl-HNO_3	– (Ashing)	NAA, XRF, Phot.	[570–573]
Cu	Ag	HCl-HNO_3	1.5 M HNO_3	Phot.	[574]
Cr	Co, Cu, Fe, Sn, Zn	HCl	HCl, HNO_3	Phot., Polar.	[575, 576]
Pb	Sb	9 M HCl	0.5–1 M NaOH	Phot.	[577]
Al, Na, S	Cl	NaOH soln.	0.5 M NH_4NO_3	Phot.	[578, 579]
Chelating resins					
Na	Cu, Mn	NaCl soln.	1 M H_2SO_4, 2 M HCl	Phot.	[580]

Table 35. Removal of matrix elements by sorption on ion exchange resins

Matrices	Trace elements	Sorbed from	Determination techniques	Lit.
Cation exchange resins				
Pu, U	B, Si	HNO_3	OES	[581, 582]
Na	B	NaOH soln.	Phot.	[583]
Anion exchange resins				
U	Rare earths, Mn, Ni, Pb	8–9 M HCl	OES, Polar.	[584, 585]
Pu, Th, U	ca. 20 elements	8 M HNO_3	OES	[586]
Ta	ca. 20 elements	HNO_3-HF	OES, Polar.	[501, 587]
Bi	Cu	3 M HNO_3	Phot.	[588]
Ga	Cu, Ni	4 M HCl	Spot test	[589]

There are many applications of this technique in the enrichment of trace impurities at the ng/g or low μg/g level in high-purity metals and other solid samples. The sample is first converted into a solution in which the desired trace elements are sorbed strongly on the resin but the matrix elements are not. Table 34 lists some examples.

9.2.4 Removal of Matrix Elements by Sorption on Ion Exchange Resins

In this case, a column containing a relatively large amount of resin is required to prevent the breakthrough of the matrix elements from the column. The operation is quite simple, but the concentration of the desired elements in the effluent is decreased by washing of the column, so that concentration of the effluent by evaporation, etc. is often required before the trace determination. Several applications of this technique in trace analyses of high-purity metals, alloys and compounds are given in Table 35.

9.2.5 Sorption of Matrix and Trace Elements on Ion Exchange Resins Followed by Chromatographic Elution

This is the most unfavorable case, where a relatively long column and careful manipulation are required to sorb the desired trace elements quantitatively and separate them by chromatographic elution from the matrix element. The proper choice of eluents is essential to achieve a successful separation. Table 36 lists some examples in trace analyses of metals.

9.3 Separation with Cellulosic Exchangers

Cellulose as such is used in paper and liquid-liquid partition chromatographies and also serves as support for various organic and inorganic solid sorbents. It is possible to introduce various functional' (cationic, anionic or chelating) groups into cellulose in

Table 36. Sorption of matrix and trace elements on ion exchange resins followed by chromatographic elution

Matrices	Trace elements	Sorbed from	Desorbed with	Determination techniques	Lit.
Cation exchange resins					
Cu, Zn	Cd	Dil. HNO_3	0.5 M HCl	Phot.	[590]
Cd	Zn	Dil. HNO_3	0.5 M HCl (Cd) $\rightarrow$ 2 M HCl (Zn)	Phot.	[590]
Th	Rare earths	–	0.5 M H_2SO_4 + 0.12 M $(NH_4)_2SO_4$ (Th) $\rightarrow$ 6.5 M HNO_3 (Rare earths)	OES	[591]
Ba, K, Na, Sr	Many	–	12.2 M HCl*	–	[592]
Al, Cr, La, Mn, Ni, Pb, Sc, Y, Alkali and alkaline earth metals	Many	–	Dioxane-HCl*, Dioxane-ethanol-HCl*	–	[593]
Anion exchange resins					
Pu, U	Many	12 M HCl	HCl, HNO_3, H_2O	OES	[594]
U	Cd	3 M HCl	1 M HCl(U) $\rightarrow$ 0.5 M HNO_3 (Cd)	Polar.	[595]
Bi	Pb	5 M HNO_3	8 M HCl	ASV	[596]

* The matrices are precipitated in the resin phase.

forms of flocs, powder and paper to convert it into cellulosic exchangers. A number of cellulosic exchangers are commercially available or rather easily prepared in analytical laboratories.

Some applications of cellulosic exchangers in trace analysis follow. For the enrichment of low $\mu g/g$ of Cd, Cu, Pb and Zn in uranium prior to the polarographic determination, the matrix is removed by sorption on cellulose phosphate from 1 M hydrochloric acid solutions [597]. Sea water acidified to pH 2.5 is passed through a p-aminobenzylcellulose column to sorb Mo at the $\mu g/l$ level, which is then eluted with 1 M ammonium carbonate solution and determined by graphite-furnace atomic absorption spectrometry [598]. Filtration through a thin layer of cellulose containing chromotropic acid as functional group offers samples for the X-ray fluorescence determination of Cu, Fe, Hg, Sr and Zn at the $\mu g/l$ or low mg/l level in water [599]. Heavy metals such as Co, Cu, Fe, Mn, Ni, Pb, Ta, U and Zn at the $\mu g/l$ level in water are sorbed on a column packed with cellulosic exchanger containing 1-(2-hydroxyphenylazo)-2-naphthol (Hyphan) as functional group and then eluted with 1 M hydrochloric acid. The heavy metals in the effluent are sorbed again on another small portion of Hyphan-cellulose at pH 7.5 by batch operation for the X-ray fluorescence determination [600, 601]. Cations including Cr^{3+}, Fe^{3+}, Co^{2+}, Ni^{2+}, Cu^{2+}, Zn^{2+}, Ag^+, Cd^{2+}, Hg^{2+}, Pb^{2+} and UO_2^{2+} and anions including VO_3^-, CrO_4^-, AsO_4^{3-}, SeO_4^{2-}, MoO_4^{2-}, WO_4^{2-} and Br^- in aqueous solutions are collected on a 10-cm^2 2,2'-diaminodiethylamine-cellulose filter disk, which is then directly used for the trace determination by X-ray fluorescence spectrometry [602, 603]. The optimum pH range is above 6 for cations and 3 to 6 for anions. Traces of Cu, Fe, Ni, Pb and Zn in various inorganic salts are separated on a column filled with cellulosic exchanger containing 4-(2-pyridylazo)-resorcinol (PAR) as functional group [604].

9.4 Separation with Polyurethane Foams

Open-cell flexible polyether-type polyurethane foams are useful as sorbents for elements which can be extracted from aqueous solutions to ethyl ether, e.g. Au(III), Fe(III) and Tl(III) from hydrochloric acid solutions, and Cd, Co(II), Fe(III) and Zn from thiocyanate solutions [605]. Polyurethane foams with immobilized organic extractants such as TBP are used in reversed-phase partition chromatography. Various kinds of polyurethane foams with immobilized chelating agents are prepared by dissolving an organic reagent (e.g. dithizone, 1-nitroso-2-naphthol, diethylammonium diethyldithiocarbamate, PAN) in a plasticizer (e.g. TBP, α-di-n-nonyl phthalate, di-n-octyl phthalate, dibutyl adipate) and then immobilizing the solution on a polyurethane foam by swelling. For example, Co in water is sorbed on a PAN-loaded polyurethane foam column, and then eluted with acetone for the spectrophotometric determination [606]. Polyurethane foams with immobilized ion exchange resins or inorganic sorbents have been prepared. Unlike all the reagent foams described above, it is possible to directly introduce specific functional groups into polyurethane foams. The foams with sulfhydryl groups are used as sorbents for mercury(II) chloride and methylmercury(II) chloride at the $\mu g/l$ or low mg/l level in aqueous solutions [607]. Polyurethane foams with ion-exchange groups have been prepared.

Reviews are available on this subject [472, 473].

9.5 Separation with Miscellaneous Organic Sorbents

Dextran gel. Traces of B and V in natural waters and rocks are sorbed on a Sephadex G-25 gel column and eluted with 0.02 M or 0.12 M hydrochloric acid for spectrophotometric determinations [608].

Preformed organic reagent precipitates. A small amount of preformed voluminous *p*-dimethylaminobenzylidenerhodanine precipitate is added to a sample solution to selectively collect traces of Ag at acidities of 0.2 to 0.5 M, where the conventional coprecipitation technique is not applicable [609]. A few $\mu g/g$ of Ag in high-purity bismuth is rapidly separated by this technique with a recovery of greater than 95% and an enrichment factor of 10^3 to 10^4 for the spectrophotometric determination.

Powdered organic reagents. Sorption of traces of Ag, Fe and Co on dithizone or 1-nitroso-2-naphthol powder in dilute nitric acid solutions is accelerated by ultrasonic irradiation [610, 611]. This technique is successfully applied to the analysis of high-purity lead for Ag. Naphthalene powder doped with 1-(2-thiazolylazo)-2-naphthol is used to collect Ni and other heavy metals from solutions at pH 6.9 [612]. The powder containing the trace metals is dried, pelletized, and delivered to X-ray fluorescence spectrometry.

Organic reagent papers. Filter papers impregnated with water-insoluble organic reagents such as dithizone and *p*-dimethyl aminobenzylidenerhodanine are easily prepared and used in rapid collection of trace metals in aqueous solutions by filtration [613, 614]. The dithizone paper has been applied to the enrichment of traces of Ag in electrolytic copper prior to the spectrophotometric determination and preparation of thin and homogeneous samples for beta activity measurements of Bi-210.

Organic reagents supported on macroreticular resin. A cross-linked copolymer of styrene and divinylbenzene, Amberlite XAD-2 resin, coated with oxine or other organic reagents, is a useful sorbent for various heavy metals [615].

Organic reagents supported on silica gel. Water-insoluble organic chelating agents such as 2-mercaptobenzothiazole, *p*-dimethylaminobenzylidenerhodanine and 1-nitroso-2-naphthol, supported on silica gel are useful for the enrichment of Cd, Cu, Pb, Zn, Hg, Ag, Au, Pd and Co in river and sea waters [616–618]. The sorbents are easily prepared and higher flow rates can be used in column operation than with conventional chelating resins.

Silica gel or glass beads with bound chelating groups. Silica gel or controlled pore glass beads with bound chelating groups such as diamines, dithiocarbamates and oxine are useful for the enrichment of trace heavy metals in aqueous solutions [619–625]. After sorption, the sorbents can be used directly for the trace determination by X-ray fluorescence spectrometry. A 5-l sea water sample buffered at pH 5.6 is introduced into a column packed with controlled pore glass beads with ethylenediaminetriacetic acid groups to sorb Cu(II), Pb and Zn, which are then eluted with 15 ml of 1 M hydrochloric acid for atomic absorption spectrometry [622].

Reversed-phase partition chromatography. As large as 3 l of fresh water acidified to 1 M with hydrochloric acid is introduced into a column containing a macroreticular resin, Amberlite XAD-2, coated with tri-*n*-octylamine dissolved in cyclohexane to sorb Cd, which is then eluted with 1 M ammonium sulfate-0.1 M EDTA (pH 5.5) for atomic absorption spectrometry [626]. Traces of Co in aqueous solutions are sorbed on a column containing silaned nonporous glass beads (treated with dimethyldichlorosilane and octadecyltrichlorosilane) as colored 2-(2-pyridylazo)-5-diethylaminophenol complex, and eluted with a 1 : 3 (v/v) mixture of ethanol and 1 M hydrochloric acid for spectrophotometry [627]. Similarly, traces of P in river and sea waters are sorbed as phosphomolybdenum blue complex and then eluted with *N,N*-dimethylformamide, while monitoring the absorbance at 700 nm of the effluent [628]. Trace heavy metals including Cd, Co, Cu, Fe, Mn, Ni, Pb and Zn in sea water are sorbed on a column packed with C_{18} chemically bonded silica gel as oxinate, and then eluted with methanol for inductively coupled plasma-optical emission spectrometry and atomic absorption spectrometry [629, 630]. Low μg/g of Ag and Cu in high-purity lead are enriched by reversed-phase partition chromatography using dithizone dissolved in *o*-dichlorobenzene on a diatomite support [631]. For the spectrographic determination of nanogram quantities of Be and rare earths in purified uranium compounds, the matrix is removed by sorption on a tri-*n*-octylamine-silica gel column from 8M hydrochloric acid solutions [632].

9.6 Separation with Activated Carbon

Table 37 shows some examples of this technique. Typically, various trace elements are quantitatively collected on about 50 mg of activated carbon from about 200 ml of sample solution by batch operation or filtration through a thin layer of the sorbent supported on a filter paper, generally in the presence of a chelating agent. The trace elements are then desorbed with nitric acid or by heating (for Hg). By proper selection of the chelating agent, recoveries of greater than 95% and enrichment factors of 10^3 to 10^4 are achieved by this technique in the enrichment of impurities at the ng/g or low μg/g level in high-purity metals and compounds. Washing commercial activated carbon with 48% hydrofluoric acid followed by 12 M hydrochloric acid before use is effective to remove such impurities as Al, Fe, K, Ti and Zn.

Table 37. Sorption of trace elements on activated carbon

Matrices	Trace elements	Chelating agents	Determination techniques	Lit.
Water	Ag, Bi, Cd, Cu, In, Mg, Mn, Pb	–	AAS	[633]
Water	Hg, Methylmercury	–	AAS	[634]
NaClO$_4$	Ag, Bi, Cd, Co, Cu, Fe, Hg, In, Mn, Ni, Pb	–	AAS	[635]
Water	ca. 20 elements	Oxine	AAS, XRF, NAA	[636]
Alkali and alkaline earth salts	Ag, Bi, Cd, Co, Cu, Fe, In, Ni, Pb, Tl, Zn	DDTC	AAS	[637]
Cr	Ag, Bi, Cd, Co, Cu, In, Ni, Pb, Tl, Zn	Hexamethylene-ammonium hexamethylene-dithiocarbamate	AAS	[638]
Mn	Bi, Cd, Co, Cu, Fe, In, Ni, Pb, Tl, Zn	Potassium xanthate	AAS	[639]
Mg	Ag, Cd, Co, Cu, Ni, Pb, Zn	Dithizone	AAS	[633]
Ag, Tl	Bi, Co, Cu, Fe, In, Pb	Xylenol orange	AAS	[640]
Al, Cr, Fe, Ga, Mn	Bi, Cd, Cu, In, Pb, Tl	Dithiophosphoric acid-*O, O*-diethyl-ester	AAS	[641, 642]

9.7 Inorganic Ion Exchangers

Inorganic ion exchangers include hydrated oxides, acidic salts of multivalent metals (e.g. zirconium phosphate), salts of heteropolyacids (e.g. ammonium molybdo-phosphate), insoluble ferrocyanides, synthetic aluminosilicates, sulfides, and alkaline earth sulfates [643, 644]. Their applications in trace analysis are limited because of their poorer reproducibility of sorption properties, lower exchange capacity and higher solubility, compared with ion exchange resins, though their resistivity to high temperatures, radiation, organic solvents and oxidizing agents, and their particular selectivity to some ions are sometimes useful.

Some examples are given below. Traces of uranyl ions are selectively sorbed on a silica gel column from solutions containing EDTA and tartaric acid at pH 5, and then eluted with 3 M acetic acid [645]. The method is applied to the spectrophotometric determination of U in rocks and underground waters. Trace heavy metals such as Co, Cu, Fe, Ni, Pb, U and Zn in fresh and sea waters are sorbed on a column packed with hydrated titanium oxide, hydrated zirconium oxide or aluminum oxide [646].

Batch operation is widely used. Sorption of traces of Sb(V) on aluminum oxide from dilute nitric acid solutions is accelerated by ultrasonic irradiation [647]. Hydrated iron(III) oxide is used for the enrichment of Cr and V at the low $\mu g/g$ level in high-purity aluminum prior to the determination by coulometric titration [648]. Traces of phosphate and arsenate ions are quantitatively collected on zinc oxide powder, which is then dissolved in 6 M hydrochloric acid [649]. This technique is applied to the spectrophotometric determination of P in water samples and of P and As in high-purity lead. Bismuth at the ng/g level in copper is selectively sorbed on hydrated lead oxide, which is then dissolved in a sodium oxalate solution for the polarographic determination of Bi [650]. Traces of As and P in dilute aqueous solutions are collected on ignited barium and strontium sulfates [651, 652]. A few $\mu g/g$ of Se in copper is sorbed on lead sulfate, which is then dissolved in an ammonium tartrate solution for spectrophotometry [397]. As little as 0.01 $\mu g/l$ of Cs in water is sorbed on ammonium 12-molybdophosphate, which is then dissolved in sodium hydroxide solution. After extraction with an MIBK-cyclohexane solution of sodium tetraphenylboron, the Cs is determined by flame photometry [653].

An aqueous sample solution (0.1 to 6 l, pH 3 to 6) is filtered through a thin (300 to 400 nm) layer of freshly prepared metal sulfides (e.g. ZnS, MnS, CuS, PbS) supported on a membrane filter to collect traces of heavy metals including Ag, Bi, Cd, Cu, Hg, Pb and Te. The trace metals are then determined by X-ray fluorescence spectrometry directly on the filter or by atomic absorption spectrometry after dissolution of the precipitate [654]. Traces of Ag, Cu and Hg in natural and waste waters are collected on a disk of filter paper impregnated with zinc sulfide, which is then directly used for the trace determination by X-ray fluorescence [655].

A tungsten, molybdenum, tantalum or rhenium wire is immersed in aqueous solutions to concentrate trace heavy metals on it by ion exchange on the oxidized surface or other mechanisms [656, 657]. The wire is then delivered to flameless atomic absorption spectrometry.

10 Flotation

Flotation is defined as the process by which suspended matter and solutes in aqueous solutions are selectively floated to the solution surface with the aid of a rising stream of gas bubbles. Hydrophobic substances are easily attached to the bubbles and floated. Therefore, when hydrophilic substances are to be floated, they are generally rendered hydrophobic in combination with suitable surfactants beforehand. A well-known industrial application of this technique is the concentration of valuable minerals in ores. Although flotation of various substances had been extensively studied [658—660], its utility as an enrichment technique in inorganic trace analysis was proved quite recently [661, 662]. Two reviews describe the application of flotation in inorganic trace analysis [663, 664].

10.1 General Procedures

Typical flotation cells are shown in Fig. 31. A rising stream of numerous tiny bubbles of air or nitrogen is produced by passing the gases through a fine-porosity sintered-

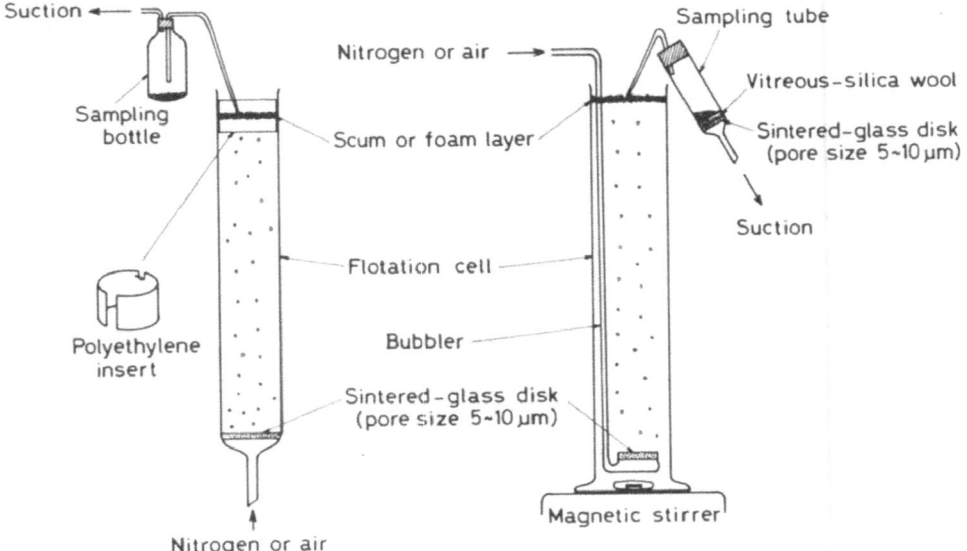

Fig. 31. Flotation cells and ancillaries

glass disk, and a scum or foam layer containing the desired substances is formed on the solution surface. The scum or foam layer is then collected with a spatula, pipet, sampling bottle or sampling tube. When the scum adheres strongly on the wall of the cell near the solution surface, a detachable polyethylene insert is helpful to completely collect it. Another method for separating the floated substances from the solution is rapid filtration by suction through the sintered-glass disk at the bottom of the cell. The copious foam, which reduces enrichment factors, can be easily destroyed by addition of small amounts of organic solvents such as ethanol, 1-butanol and ethyl ether, or by contact with vapors of the above solvents.

10.2 Carrier Precipitation Followed by Flotation

In this technique, desired trace elements in an aqueous sample solution are quantitatively collected on small amounts of inorganic or organic collector precipitates, which are then floated by bubbling with or without the aid of surfactant ions of opposite charge to the precipitate surfaces.

10.2.1 Important Experimental Factors

Collector precipitates. These should be selected from both standpoints of carrier precipitation (see Chap. 7.2) and flotation. In general, bulky flocculent precipitates larger than gas bubble diameters are desirable for successful flotation, because numerous tiny gas bubbles are easily trapped in the interstitial spaces and on the surfaces of the precipitates to give sufficient buoyancy. Therefore, collector precipitates are coagulated by mechanical stirring of the sample solution before the flotation. Table 38 lists collector precipitates suitable for flotation [665, 666]. Usually, 10 to 100 mg of collector precipitates are used for 100 to 1 000 ml of sample solutions.

Bubbling. A sintered-glass disk (5- to 10-μm nominal pore size) is used. Gas bubbles having diameters of 0.1 to 0.5 mm are desirable, because they are easily trapped by flocculent precipitates and also form a stable foam layer on the solution

Table 38. Floatable collector precipitates

Inorganic
$Fe(OH)_3$, $Al(OH)_3$, $Cr(OH)_3$, $Ti(OH)_4$, $Zr(OH)_4$, $Mg(OH)_2$, $Sn(OH)_4$, $Bi(OH)_3$, $In(OH)_3$, $Fe(OH)_2$, $Co(OH)_2$, $Ni(OH)_2$, $Cu(OH)_2$, $Zn(OH)_2$, $Sb(OH)_3$, $Th(OH)_4$, CdS, PbS

Organic
Thionalide,
Dithizone,
p-Dimethylaminobenzylidenerhodanine,
1-Nitroso-2-naphthol,
2-Mercaptobenzimidazole,
2-Mercaptobenzothiazole,
α-Benzoin oxime

---

surface in the presence of surfactants. Addition of small amounts (about 1%) of organic solvents such as methanol, ethanol, acetone and methyl cellosolve is essential to obtain a stream of numerous tiny bubbles, because these solvents prevent coalescence of tiny gas bubbles which appear from adjacent pores of a sintered-glass disk.

The gas flow rate is adjusted to stir the solution gently so that the gas bubbles collide against the precipitates frequently. Flow rates of 1 to 2 ml cm^{-2} min^{-1} are generally used for solution volumes of 100 to 3000 ml. The required bubbling time varies between several seconds and several minutes.

pH of the solution. The optimum pH range is determined mainly from the standpoint of trace recoveries in carrier precipitation. Carrier precipitation and flotation are generally carried out at the same pH.

Surfactants. In general, inorganic collector precipitates need surfactants for flotation. Surfactant ions of opposite charge to precipitate surfaces are used to turn the hydrophilic surfaces to hydrophobic. Since the surface charge of the hydroxide precipitates depends on the pH of the solution and changes its sign at the isoelectric point, either cationic or anionic surfactants should be effective at a given pH. Frequently, however, both cationic and anionic surfactants are effective over a relatively wide pH range as shown in Fig. 32 [667]. This is probably due to the fact that flocculent precipitates can trap tiny bubbles in their interstitial spaces in spite of their hydrophilic slightly charged surfaces.

Another important role of the surfactants is to form a stable foam layer to support the floated precipitates on the solution surfaces, which is often important to quantitatively collect the precipitates.

Anionic surfactants, sodium oleate and sodium dodecyl sulfate, are most frequently used. Surfactants are generally dissolved in ethanol, which is effective to produce tiny gas bubbles as described previously. Combined use of two kinds of anionic surfactants, sodium oleate and sodium dodecyl sulfate, is recommended in the flotation of flocculent indium hydroxide precipitates in sea water [668]. In this case, the two surfactants function separately, i.e. the former in the flotation and the latter in the formation of the stable foam layer on the solution surface.

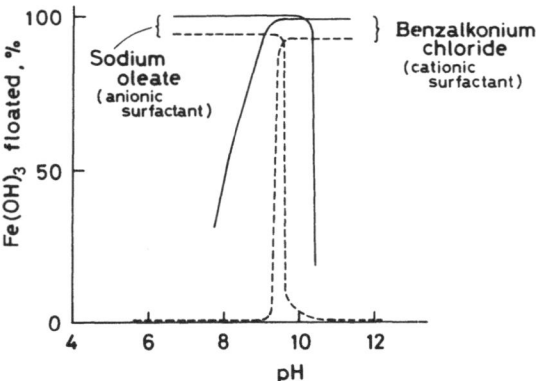

Fig. 32. Flotation of iron(III) hydroxide precipitates. In the presence (———) or absence (– – – –) of ethanol

Flotation of inorganic gathering precipitates *without surfactants* is possible in some cases. For example, bulky flocculent iron(III) hydroxide precipitates are floated by bubbling with small amounts (about 1%) of methyl cellosolve [669], or with solid paraffin (melting point: 56 to 58 °C) particles having diameters below 1 mm produced by adding a hot (about 65 °C) ethanol solution of paraffin to an aqueous sample solution [670].

Both without and with surfactants, bulky flocculent organic collector precipitates are easily floated with the aid of tiny gas bubbles and stably supported on the solution surface. Overuse of surfactants should be avoided, because the resulting copious foam can not be destroyed with organic solvents without redissolution of the collector precipitates.

10.2.2 Applications

This enrichment technique has been extensively applied to water analysis (Table 39). Various trace elements at the $\mu g/l$ or ng/l level in 250 to 3000 ml of fresh and sea waters are quantitatively concentrated with enrichment factors of several tens to several hundreds. For sea water, most of alkali and alkaline earth elements are removed. The technique using indium hydroxide collector precipitates can be applied to the simultaneous enrichment of trace heavy metals, which are adsorbed on and

Table 39. Enrichment of trace elements in water samples by carrier precipitation followed by flotation

Trace elements	pH	Collector precipitates	Determination techniques	Lit.
As(V), Bi, Mo(VI), P(V), Sb(III, V), Se(IV), Sn(II,IV)	4	$Fe(OH)_3$	AAS, Phot.	[671–676]
V(V)	5	$Fe(OH)_3$	AAS	[677]
Se(IV)	3.5–5.3	$Fe(OH)_3$	Phot.	[678]
U(VI)	5.7	$Fe(OH)_3$	Phot.	[679]
Cu(II), Zn	7.6	$Fe(OH)_3$	AAS, Phot.	[680]
As(III,V)	8–9	$Fe(OH)_3$	AAS	[681, 682]
Cd, Co, Cr(III), Cu(II), Fe(III), Mn(II), Ni, Pb, Zn	9.5	$Al(OH)_3$	AAS	[683]
Cd, Co, Cr(III), Cu(II), Mn(II), Ni, Pb	9.5	$In(OH)_3$	ICP-OES	[668]
U(VI)	5.7	$Th(OH)_4$	Fluor.	[684]
U(VI)	6.6	$TiO_2 \cdot xH_2O*$	Phot.	[685]
Hg(II)	1	CdS	AAS	[686]
Ag	2	PbS	AAS	[687]
Ag	1	2-Mercapto-benzothiazole	AAS	[688]

*Preformed precipitate

Table 40. Enrichment of trace impurities in metals by carrier precipitation followed by flotation

Matrices	Trace elements	pH	Collector precipitates	Determination techniques	Lit.
Zn	Sn(IV)	6	$Fe(OH)_3$	Phot.	[670]
Zn	Fe(III), Pb	Ammoniacal soln.	$Bi(OH)_3$	AAS	[690]
Pb, Zn	Ag, Cu(II)	1–1.5	Dithizone	AAS	[691]
Cu	Ag	1	p-Dimethyl-amino-benzylidene-rhodanine	Phot.	[661]
Zn	Co	3	1-Nitroso-2-naphthol	AAS	[692]

occluded in suspended particles, complexed with humic acid, and existing as inorganic colloids and ions, in fresh waters [689].

Trace impurities at the ng/g or low μg/g level in pure metals are also enriched by carrier precipitation followed by flotation (Table 40).

The advantages of this technique are:

(1) This technique is more rapid and convenient than conventional carrier precipitation techniques using time-consuming and troublesome filtration and centrifugation for the separation of collector precipitates from mother liquor.

(2) Flotation of bulky flocculent precipitates needs less experimental skill and is more reliable and rapid than ion flotation described in Chap. 10.3.

(3) Generally, much higher enrichment factors are obtained than by ion flotation.

10.3 Ion Flotation

In this technique, desired trace ions in an aqueous solution are converted into hydrophobic compounds by reactions with surfactants and complexing agents, floated by bubbling, and concentrated in a scum or copious foam layer on the solution surface. Ion flotation in which the floated substances are dissolved in a water-immiscible organic solvent placed on the solution surface is also useful (solvent sublation).

10.3.1 Important Experimental Factors

Surfactants. A surfactant which is of opposite charge to the desired ions or their complex ions and reacts selectively with them is used. The quantity of the surfactant should be a little greater than the stoichiometric amount, but the more excess may decrease the trace recoveries.

pH of the solution. The optimum pH range should be selected from the standpoint of the reactions of the desired ions with the complexing agent and the surfactant.

Bubbling. A rising stream of gas bubbles is produced by using a sintered-glass disk (20- to 30- or 5- to 10-μm nominal pore size). Bubbling time and gas flow rate should be carefully optimized in each case, because it is generally difficult to observe when the flotation is complete.

Coexisting ions. An increase in the quantity of coexisting other ions generally decreases the trace recoveries, probably because of competition between the desired ions and the other ions for surfactants.

10.3.2 Applications

Applications of this technique to trace constituents at the μg/l level in fresh and sea waters are summarized in Table 41.

In the enrichment of 0.1 to 1 μg of Ag, Au(III), Co, Cu(II) or Fe(III) from 200 ml of aqueous sample solutions containing 0.5 to 3 g of Mg, Na or Zn, ion flotation was repeated twice to improve enrichment factors up to about 1000 with trace recoveries of greater than 90% [662].

Table 41. Enrichment of trace elements in water samples by ion flotation

Trace elements	Complexing agent and surfactant	Determination techniques	Lit.
U(VI)	Arsenazo III, Tetradecyldimethyl-benzylammonium chloride	Phot.	[693, 694]
Cr(VI)	Cetylethyldimethylammonium bromide	Phot.	[695]
Cr(VI)	Diphenylcarbazide, Sodium dodecyl sulfate	Phot.	[696]
Cu(II), Methyl-mercury(II) chloride	Potassium n-butyl xanthate, Cetyltrimethyl-ammonium bromide	AAS, Phot., Gas chromato-graphy	[697–699]
Sulfide	N, N-dimethyl-p-phenylene-diamine, Sodium dodecyl sulfate	Phot.	[700]
Nitrite	p-Aminobenzenesulfonamide, N-1-naphthylethylenediamine, Sodium dodecyl sulfate	Phot.	[701]
Cu(II)*, Fe(III)*	3-(2-Pyridyl)-5,6-diphenyl-1,2,4-triazine, Sodium dodecyl sulfate	Phot.	[702, 703]

*Solvent sublation

11 Freezing and Zone Melting

The enrichment techniques described in this chapter are based on the segregation, a phenomenon that the impurity concentration (C_S) in the just-freezing solid differs from that (C_L) in the liquid. A typical impurity concentration profile in normal freezing is shown in Fig. 33. The effective distribution coefficient, k ($\equiv C_S/C_L$), determines how effectively the impurity is concentrated; the more the k value differs from unity, the larger the enrichment efficiency becomes, and, at k = 1, enrichment is not effected at all. This coefficient is expressed by the following equation.

$$k = \frac{K}{K + (1 - K) \exp (-f\delta/D)} \tag{46}$$

where K is the equilibrium distribution coefficient, i.e. the ratio of solidus and liquidus concentrations at a given temperature in a phase diagram, f is the growth rate of the solid, δ is the thickness of the diffusion layer, and D is the diffusion constant of the impurity in the liquid. Thus, k varies from unity to K with decreasing growth rate, increasing diffusion constant, and increasing degree of mixing of the liquid.

11.1 Freeze Concentration of Dilute Aqueous Solutions

It is possible to concentrate dilute aqueous solutions by removing most of the water as pure ice [704—712]. This is a special case of normal freezing where k is virtually

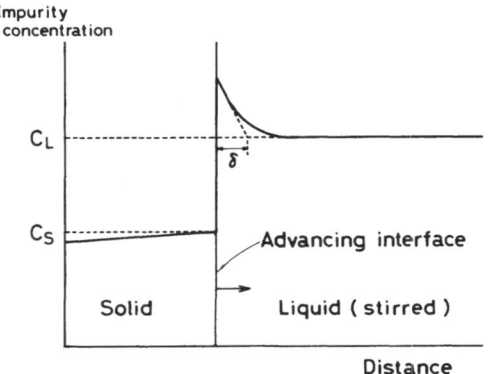

Fig. 33. Impurity concentration profile in normal freezing (k < 1)

zero. Solutions of several ten milliliters to several liters are concentrated about 10- to 100-fold in a glass or polyethylene vessel by this technique. Such trace elements as Hg at the μg/l level are recovered in greater than 95% yields when clear ice is produced under proper conditions. Freeze concentration, though time-consuming, has an advantage that losses of trace elements due to volatilization or chemical reactions are minimized.

11.2 Enrichment of Impurities in Solids by Zone Melting

In zone melting, a sequence of short molten zones are passed slowly through a relatively long solid sample in one direction. The sample is held horizontally or vertically in an appropriate container. To eliminate contamination due to the container, the floating-zone technique is used, where a molten zone is held in place by its surface tension as shown in Fig. 34 without the container.

Impurities become concentrated near the end of the sample when k < 1, and at the beginning of the sample when k > 1. The distribution of impurities in the sample after zone melting is calculated by mathematical and computational techniques [713]. Examples of impurity distribution curves are shown in Fig. 35. After many zone passes the impurity distribution reaches a steady state called ultimate distribution. The single-pass distribution (n = 1) and the ultimate distribution (n = ∞) are analytically described by Eqs. (47) and (48), respectively.

$$
\begin{aligned}
C/C_0 &= 1 - (1 - k)\exp(-kx/l) && (x \leqslant L - l) \\
&= \left\{ (L - x)/l \right\}^{k-1} && (L - l \leqslant x < L)
\end{aligned}
\tag{47}
$$

$$
C/C_0 = A\exp(Bx) \qquad\qquad (x < L - l) \tag{48}
$$

where C is impurity concentration, C_0 is initial impurity concentration, x is distance from the beginning of the sample, L is sample length, l is molten-zone length, and A and B are constants obtainable from

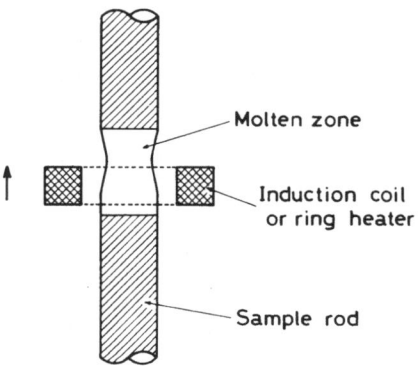

Fig. 34. Floating-zone technique

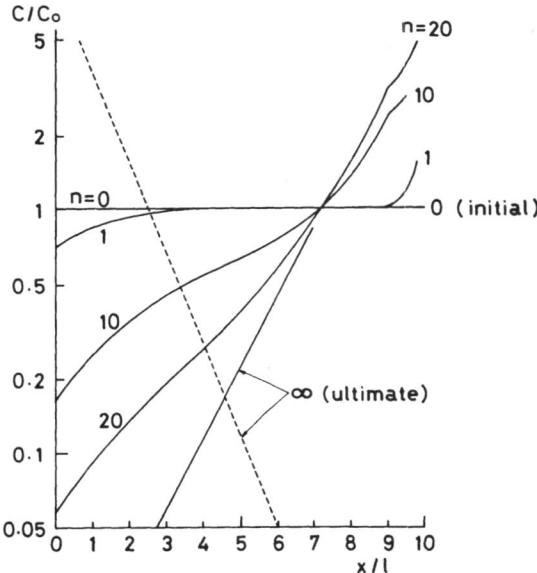

Fig. 35. Impurity distribution curves for various numbers of passes n [713]. $L/l = 10$, (————) $k = 0.7$, (– – – –) $k = 1.5$

$$k = Bl \left\{ \exp(Bl) - 1 \right\}^{-1} \tag{49}$$

$$A = BL \left\{ \exp(BL) - 1 \right\}^{-1} \tag{50}$$

The narrower the zone length is, the steeper the gradient of ultimate distribution curve (in the semilog plots) becomes.

Zone-melting, a well-known technique for purification of high-purity materials, is also useful as an enrichment technique in inorganic trace analysis. An example is the application of the floating-zone technique to the determination of boron and phosphorus in high-purity silicon by electrical conductivity measurements [714]. Because phosphorus is a donor and boron is an accepter, the conductivity is proportional to the difference in their concentrations. Fortunately, the two elements have different k values, i.e. 0.4 for phosphorus and 0.8 for boron. After 40 zone passes, the phosphorus concentration is reduced to a negligible level, and the boron only remains giving the p-type conductivity. Thus, the original concentrations of the two elements are calculated from the two conductivity measurements before and after the zone melting.

12 Enrichment Techniques in Water Analysis

In natural and waste waters, each trace element may exist in various chemical forms. For example, in fresh and sea waters, trace metal elements may exist as simple hydrated ions of different oxidation states, inorganic and organic complex ions, nonionic dissolved species, and colloids. Trace metals are also frequently adsorbed on, occluded in, or included in, inorganic, organic or biological suspended particulate matter [715−719]. Information on the chemical states of trace elements, or trace element speciation, in natural and waste waters is highly required in studies of geochemistry, environmental problems, biological effects of trace metals and water treatment. In some cases, it is possible to calculate equilibrium concentrations of various species by using published equilibrium data, pH, oxidation-reduction potentials, etc. Such techniques as potentiometric measurements with ion-selective electrodes, voltammetry and spectrophotometry, can sometimes be used to directly determine a trace element in a specified chemical form. Applicability of these techniques, however, is rather limited. Selective enrichment techniques for a specified chemical form of a trace element complement these in trace element speciation in natural and waste waters.

Various techniques are described in previous chapters for concentrating trace elements in water samples. Care must be taken, however, in direct applications to natural and waste waters of enrichment techniques such as liquid-liquid extraction and ion exchange, which have been tested only on synthetic ionic solutions, and in interpretation of the results obtained [720−722]. To obtain the total concentrations of trace metals in natural and waste waters, the following pretreatments are generally required:

(1) separation of suspended particulate matter by filtration or centrifugation

(2) destruction of organic matter by wet or dry oxidation or ultraviolet irradiation

(3) alteration of oxidation states of ions by oxidation or reduction.

The separated suspended particulate matter is analyzed for the trace metals after destruction.

This chapter describes enrichment techniques used for trace element speciation. In the applications of these techniques, it must be kept in mind that equilibria between various species may be shifted upon removal of a specified species.

12.1 Separation Based on the Particle Size and Density

12.1.1 Filtration and Ultrafiltration

Filtration is most widely used to separate suspended particulate matter in natural waters [723]. The 0.45-μm pore diameter membrane filter is commonly used to define the suspended particulate matter and the dissolved or soluble fraction. Filter materials include cellulose, synthetic polymers, glasses and metals. Filters of different materials and nominal pore diameters are commercially available. For samples with a high content of suspended matter, clogging of pores may reduce the pore diameter and retard the filtration. Stirred vacuum or pressure filtration and frequent replacement of filters are useful to overcome this difficulty. Other problems in filtration include adsorption of solutes on filters, contamination due to impurities in filter materials, airborne contamination, and rupture of phytoplankton cells by vacuum or pressure filtration.

Ultrafiltration with a Diaflo Ultrafiltration Membrane UM-2 (cut-off molecular weight: about 1000) is used to concentrate trace elements associated with humus in lake waters [724].

12.1.2 Dialysis

Dialysis is carried out on filtered sea water samples placed in a cellulose dialysis tubing of 4.8 nm mean pore diameter [725]. The distilled water is changed periodically and the dialysis is continued until a negative test for chloride is obtained. A considerable part (up to about 30%) of zinc is non-dialyzable, presumably because it is associated with organic matter. In an *in situ* dialysis technique, a dialysis bag (4.8 nm mean pore diameter) filled with pure water is directly immersed into natural water *in situ* to separate truly dissolved forms of trace elements without substantial influence of adsorption on the walls of vessels and apparatus [726]. This technique is also employed in an investigation of the interaction between humus and trace element in lake water [724].

12.1.3 Gel Filtration

Gel filtration chromatography is based on inclusion and subsequent elution of solutes through a column containing a porous polymeric gel as molecular sieve. This technique is used to fractionate dissolved organometallic compounds such as trace metal elements associated with humic materials in natural and waste waters according to their molecular size or molecular weight differences [727–730]. For example, two distinct peaks are observed in a chromatogram for a secondary sewage effluent spiked with 700 ng of Cu. The larger molecular weight fractions (molecular weight: about 10,000) contain about 100 ng of Cu, while the smaller molecular weight fractions (molecular weight: 500 to 1000) contain about 700 ng of Cu. No free copper is found in the sample. This technique is also applied to the evaluation of organo-metallic interactions in natural waters.

12.1.4 Centrifugation

Centrifugation at 48,000 G is applied to sea water [731]. From the experimental results, it seems probable that part of Cd, Cu, Pb and Zn are associated with the colloidal and fine particulate matter.

12.2 Separation Based on Chemical Reactivity

12.2.1 Volatilization

This technique is useful in speciation of arsenic, selenium, tin, antimony and mercury at the ng/l or μg/l level in natural waters.

As. Volatile arsines such as mono-, di- and trimethylarsines can be separated from other nonvolatile arsenic species by simple gas stripping, i.e. bubbling a helium stream through the sample [732]. Then, inorganic As(III) is selectively converted into arsine with sodium borohydride at pH 4 to 6 and stripped from the solution [732–735]. Inorganic As(III) plus As(V), and methylarsenic, dimethylarsenic and trimethylarsenic compounds are volatilized by reducing them simultaneously at pH 0 to 1 with sodium borohydride to arsine, methylarsine, dimethylarsine and trimethylarsine, respectively. The arsines are collected in a liquid nitrogen cold trap, then separated from each other either by fractional volatilization or by gas chromatography, and detected by atomic absorption spectrometry or optical emission spectrometry or with electron capture and flame ionization detectors [732, 736–738].

Se. Volatile dimethyl selenide and dimethyl diselenide can be separated from other nonvolatile selenium species by simple gas stripping. They are collected in a cold trap, and then separated from each other by gas chromatography. The differentiation of nonvolatile inorganic Se(IV) and Se(VI) is carried out by selectively volatilizing Se(IV) as hydride from 4 M hydrochloric acid solutions with sodium borohydride. Se(VI) is quantitatively reduced to Se(IV) by simply boiling an acidified solution for obtaining the total inorganic selenium value. All the selenium species are determined by atomic absorption spectrometry [739].

Sn. Inorganic Sn(IV) and the halides of methyltin, dimethyltin, trimethyltin, diethyltin, triethyltin, *n*-butyltin, di-*n*-butyltin, tri-*n*-butyltin and phenyltin in natural waters are volatilized as hydrides with sodium borohydride. The hydrides are then separated from each other on the basis of their different boiling points and detected by atomic absorption spectrometry [740].

Sn. Inorganic Sn(IV) and the halides of methyltin, dimethyltin, trimethyltin, diare volatilized as stibine, methylstibine, and dimethylstibine by reduction with sodium borohydride, collected on a cold trap, separated chromatographically, and determined by atomic absorption spectrometry. Antimony(III) is selectively reduced at near-neutral pH, where no reduction of Sb(V) occurs [741].

Hg. Under proper conditions, inorganic mercury is selectively reduced to metallic state with tin(II) solutions [742–745] or sodium borohydride solutions [746] in the presence of organomercury compounds and volatilized from the solution for cold-vapor atomic absorption spectrometry. Total mercury is obtained by simultaneous reduction of both inorganic and organic mercury under other conditions. Differentiation among three species of mercury, i.e. inorganic mercury, arylmercury compounds such as phenylmercury(II) chloride, and alkylmercury compounds such as methylmercury(II) chloride, can be achieved by changing the reducing agents [747]. EDTA plus hydroxyl-amine reduces only inorganic mercury to elemental mercury in an alkaline solution. EDTA plus tin(II) chloride reduces inorganic mercury and arylmercury compounds to elemental mercury. Cadmium chloride plus tin(II) chloride reduces all forms of mercury to elemental mercury. An automated system is developed based on this enrichment technique and cold-vapor atomic absorption spectrometry.

12.2.2 Liquid-Liquid Extraction

Nonpolar organic species in natural waters are extracted with organic solvents. Part of copper in sea water is extracted with pure chloroform, suggesting the presence of the copper-organic complexes [748]. Part of Cu, Mn, Pb or Zn in sea water is not extract-ed with DDTC-chloroform or APDC-MIBK, presumably due to their association with particulates or colloids or their presence as strongly-bound complexes [725, 749].

Liquid-liquid extraction is also employed in the differentiation of trace elements in different oxidation states, e.g. Cr(III)-Cr(VI), As(III)-As(V) and Se(IV)-Se(VI) in natural waters. After the isolation of the trace element in a specified oxidation state, it is determined by atomic absorption spectrometry, isotope dilution-mass spectro-metry, gas chromatography, etc. Chromium(VI) is selectively extracted with DDTC-MIBK or APDC-MIBK (or -chloroform), leaving Cr(III) in the aqueous phase [750–752]. Chromium(VI) plus (III) is obtained either by the extraction after oxidation of Cr(III) to (VI) with cerium(IV), persulfate or permanganate, or by the simultaneous extraction of Cr(VI) and Cr(III) under proper conditions. Aliquat-336 (high-molec-ular-weight ammonium salt)-toluene is used for the selective extraction of Cr(VI) and Cr(III) from weakly acidic (pH 2) solutions and from neutral (pH 6 to 8) solutions containing at least 1 M thiocyanate, respectively [753]. Arsenic(III) is selectively extracted with APDC-MIBK (or -nitrobenzene) [754, 755] or ammonium sec-butyl-dithiophosphate-hexane [756], leaving As(V) in the aqueous phase. Arsenic(III) plus (V) is obtained either by the extraction after reduction of As(V) to (III) with potas-sium iodide or sodium hydrogensulfite-sodium thiosulfate, or by the simultaneous extraction of As(III) and As(V) under proper conditions. Selenium(IV) is selectively extracted with DDTC into carbon tetrachloride [757], or with 1,2-diamino-3,5-di-bromobenzene [758] or 4-nitro-o-phenylenediamine into toluene as piazselenol [759], leaving Se(VI) in the aqueous phase. Total selenium is obtained by the extraction either after conversion into Se(IV) or after reproducible proportionation of all selen-ium species between Se(IV) and Se(VI) by photo-oxidation.

12.2.3 Carrier Precipitation

This technique is used to *differentiate oxidation states of chromium and selenium* in natural waters. Chromium(III) is coprecipitated with iron(III) hydroxide precipitates, leaving Cr(VI) in solution [760, 761]. Chromium(VI) is coprecipitated either with barium sulfate precipitates after masking of Al, Cr(III) and Fe(III) with salicylic acid [762], or with cobalt pyrrolidinedithiocarbamate precipitates at pH 4.0 after removal of Cr(III) by carrier precipitation with iron(III) hydroxide at pH 8.5 [761]. The chromium is determined by spectrophotometry or X-ray fluorescence spectrometry. Selenium(IV) is coprecipitated with iron(III) hydroxide precipitates, and Se(VI) remaining in solution is collected on elemental tellurium precipitates produced by reduction of Te(IV) with hydrazine sulfate in an acidic solution [763]. The selenium is then determined by fluorometry.

12.2.4 Electrodeposition

At pH 4.7, both Cr(VI) and Cr(III) are reduced and accumulated as metallic chromium with mercury on a pyrolytic graphite tube cathode at −1.8 V vs. SCE. At the same pH, but at −0.3 V vs. SCE, Cr(VI) is selectively reduced to Cr(III) and accumulated by adsorption [413].

12.2.5 Sorption, Ion Exchange and Liquid Chromatography

A substantial fraction of Cd, Cu, Ni, Pb and Zn in sea water and sewage effluent is not retained by a chelating resin (Chelex 100, Dowex A-100) column, presumably adsorbed on, or occluded in, organic or inorganic colloidal and fine particulate matter, or existing as relatively stable organic complexes [731, 749, 764−766].

Organic compounds in sea water are sorbed on a column containing macroreticular resin beads having no ion-exchangeable functional groups (e.g. Amberlite XAD-2) and then eluted with methanol and aqueous ammonia. Inorganic metal ions are not sorbed on the resin. By this technique, more than 80% of Cd, Cu, Fe and V dissolved in sea water are found to be present as organic compounds [767, 768]. Methylarsenic compounds can be separated from inorganic As(III) and As(V) by ion-exchange chromatography [769, 770]. Methylmercury and Hg(II) sorbed on activated carbon are differentiated by desorption with 0.1 M nitric acid-95% acetone [771].

Ion exchange and other sorption techniques are also used to *differenciate oxidation states* of a trace element dissolved in natural and waste waters, e.g. Cr(III)-Cr(VI) and Se(IV)-Se(VI). Chromium(VI) in natural waters is sorbed at pH 5 on an anion-exchange resin column, whereas Cr(III) is not. The sorbed Cr(VI) is then eluted with 1 M sodium chloride or a reductant solution [0.5 M iron(II) ammonium sulfate in 1 M hydrochloric acid] for atomic absorption spectrometry [772]. Chromium(III) and Cr(VI) in waste waters are converted into their APDC complexes and separated from each other by reversed-phase high-performance liquid chromatography [773]. Chromium(III) is collected on preformed hydrated iron(III) oxide, leaving Cr(VI) in solution [774−776]. Chromium(III) plus Cr(VI) is collected on hydrated iron(III) oxide after reduction with sodium sulfite, or with hydrated iron(II) oxide or hydrated bismuth oxide. Chromium(VI) is selectively sorbed on a poly(dithiocarbamate) resin column [777]. Selenium(IV) is selectively reduced to elemental selenium with L-ascorbic acid in the presence of Se(VI) in water samples, and sorbed on activated carbon, which is then filtered off for X-ray fluorescence spectrometry [778]. The total selenium content is obtained after reduction by refluxing the water samples with thiourea in sulfuric acid and subsequent sorption of the resulting elemental selenium on activated carbon. Selenium(IV) is also selectively sorbed on a macroreticular resin (Amberlite XAD-2) column as Se(IV)-DDTC complex in the presence of Se(VI) in sea water [779]. The sorbed Se(IV) is then eluted with a dilute mixture of nitric and perchloric acids and determined by fluorometry.

13 Enrichment Techniques in Gas Analysis

In the trace analysis of the environmental atmosphere and other gaseous samples, enrichment of liquid and solid particles (see Fig. 2) and gaseous trace constituents at the low part per million (v/v) level or lower is a matter of common practice.

13.1 Separation of Particles

The techniques for collecting airborne or gas-borne particles are based on one of the following principles: filtration, impaction, sedimentation, centrifugation, thermal precipitation, and electrostatic precipitation [780, 781].

Both filters and impactors are most widely used. In the former, a large volume of gas sample is sucked at an appropriate flow-rate through a fiber or membrane filter having nominal pore sizes of about 1 to 10 μm. Filter materials used include vitreous silica, glasses, metals, cellulose derivatives and synthetic polymers. Impactors collect airborne particles by sucking the gas sample through a nozzle with high linear velocities and depositing the particles on a plate immediately in front of the nozzle. Several such stages are joined in series in *cascade* or *multistage impactors* (Fig. 36). Typically, the first stages have nozzle diameter of several millimeters and the last ones are

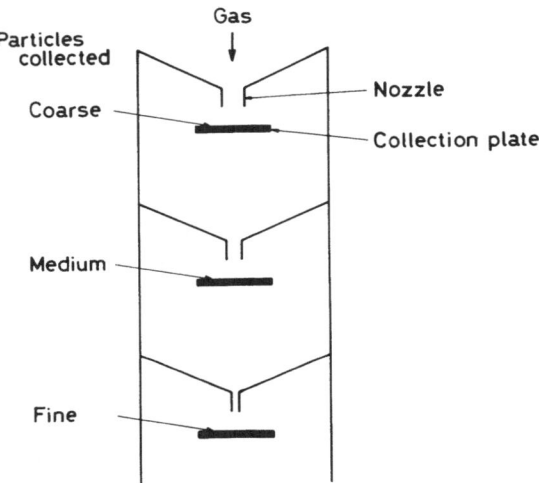

Fig. 36. Cascade impactor

several tenths of a millimeter. The Andersen sampler consists of eight stages, each having 400 nozzles. Particle size distribution between about 0.5 and 10 μm is obtained by this apparatus.

The collected particles are then observed under an optical or electron microscope, or analyzed by various optical and electrochemical determination techniques, spark source mass spectrometry, neutron activation, X-ray fluorescence and diffraction, microprobe techniques, etc. [781].

13.2 Separation of Gaseous Trace Constituents

Selective absorption of desired trace constituents in solutions is widely used. A relatively large volume of gas sample is passed through a bubbler containing a suitable absorbing solution (Table 42) at an appropriate flow-rate, tiny bubbles being produced in the solution with a fine-porosity sintered-glass disk.

Sorption on solid sorbents is also useful. Hydrogen sulfide in air is sorbed on an OH-form anion exchange resin column by suction, and eluted with 4 M sodium hydroxide solution for spectrophotometric determination by the methylene blue method [782]. Another technique uses cadmium(II)-exchanged zeolite AAA molecular sieve as sorbent, whose color turns from white to yellow (CdS) upon sorption of hydrogen sulfide [783]. The immobilized sulfide is then determined by the photometric methylene blue method. Atmospheric ammonia is collected on a tube containing porous silica beads (80 to 100 mesh) treated with potassium hydroxide, and desorbed at 280 °C for gas chromatographic determination with a chemiluminescent detector [784]. Mercury vapor in air is collected on activated carbon, silver, gold and manganese dioxide [785] for cold vapor atomic absorption spectrometry. Filter papers impregnated with reagents are also used to enrich trace constituents in gas samples.

Most of gases except for hydrogen, nitrogen, oxygen, carbon monoxide, methane and rare gases are condensed in a liquid nitrogen cold trap. Hydrogen selectively passes through a heated palladium or palladium-silver alloy tube. These techniques are sometimes useful in enrichment of trace constituents in gas samples.

Table 42. Absorbing solutions for some gaseous trace constituents

Gases	Absorbing solutions
SO_2	$HgCl_2$ + NaCl
H_2S	$CdSO_4$ + NaOH,
	$ZnSO_4$ + NaOH + $(NH_4)_2SO_4$
NO_2	NaOH
NH_3	H_3BO_3, H_2SO_4
HCN	NaOH
HF	NaOH
HCl	Water
Cl_2	o-Tolidine + HCl
Hg	$KMnO_4$ + H_2SO_4

Literature

1. Mizuike, A., Pinta, M.: Pure & Appl. Chem. *50*, 1519 (1978)
2. Doležal, J. et al.: Decomposition Techniques in Inorganic Analysis, London, Iliffe Books Ltd. 1968
3. Šulcek, Z. et al.: Crit. Rev. Anal. Chem. *6*, 255 (1977)
4. Gorsuch, T.T.: The Destruction of Organic Matter, Oxford, Pergamon 1970
5. Starik, I.E.: Osnovy radiokhimii, Moscow-Leningrad, Izd. Akademiya Nauk SSSR 1959, 1969²
6. Beneš, P., Majer, V.: Trace Chemistry of Aqueous Solutions − General Chemistry and Radiochemistry, Amsterdam-Oxford-New York, Elsevier 1980
7. Paulhamus, J.A.: Airborne Contamination, in: Ultrapurity (ed.) Zief, M., Speights, R., p. 255, New York, Marcel Dekker 1972
8. Dams, R. et al.: Anal. Chem. *42*, 861 (1970)
9. Morgan, G.B. et al.: Science *170*, 289 (1970)
10. The Japan Radioisotope Association: Environmental Monitoring by Activation Analysis, p. 105, Tokyo, 1979
11. Zoller, W.H. et al.: Science *183*, 198 (1974)
12. Zief, M., Mitchell, J.W.: Contamination Control in Trace Element Analysis, New York, Wiley 1976
13. Maienthal, E.J.: U.S. Nat. Bur. Stds. Tech. Note 545, 53 (1970)
14. Sandell, E.B., Onishi, H.: Photometric Determination of Traces of Metals − General Aspects, New York, Wiley 1978
15. Massee, R. et al.: Anal. Chim. Acta *127*, 181 (1981)
16. Kuehner, E.C., Freeman, D.H.: Containers for Pure Substances, in: Purification of Inorganic and Organic Materials (ed.) Zief. M., p. 297, New York, Marcel Dekker 1969
17. Kotz, L. et al.: Talanta *26*, 681 (1979)
18. Adams, P.B.: Glass Containers for Ultrapure Solutions, in: Ultrapurity (ed.) Zief, M., Speights, R., p. 293, New York, Marcel Dekker 1972
19. Robertson, D.E.: Contamination Problems in Trace-Element Analysis and Ultrapurification, in: Ultrapurity (ed.) Zief, M., Speights, R., p. 207, New York, Marcel Dekker 1972
20. Hetherington, G., Bell, L.W.: Vitreous Silica, in: Ultrapurity (ed.) Zief, M., Speights, R., p. 353, New York, Marcel Dekker 1972
21. Thiers, R.E.: Separation, Concentration and Contamination, in: Trace Analysis (ed.) Yoe, J.H., Koch, Jr., H.J., p. 637, New York, Wiley 1957
22. Thiers, R.E.: Contamination in Trace Element Analysis and its Control, in: Methods of Biochemical Analysis (ed.) Glick, D., Vol. 5, p. 273, New York, Interscience 1957
23. Vasilevskaya, L.S.: Working with High-Purity Materials, in: Analysis of High-purity Materials (ed.) Alimarin, I.P., p. 13, Jerusalem, Israel Program for Scientific Translations 1968
24. Mori, S.: Anal. Chim. Acta *108*, 325 (1979)
25. Bothner, M.H., Robertson, D.E.: Anal. Chem. *47*, 592 (1975)
26. Cragin, J.H.: Anal. Chim. Acta *110*, 313 (1979)
27. Thompson, G., Bankston, D.C.: Appl. Spectry. *24*, 210 (1970)
28. Ando, A.: Bunseki *1975*, 512 (1975)

29. Mizuike, A., Iino, A.: Anal. Chim. Acta *111*, 251 (1979)
30. Mizuike, A., Iino, A.: ibid. *124*, 427 (1981)
31. Mizuike A, Iino, A.: Bunseki Kagaku *27*, 358 (1978)
32. Tschöpel, P. et al.: Fresenius Z. Anal. Chem. *302*, 1 (1980)
33. Karin, R.W. et al.: Anal. Chem. *47*, 2296 (1975)
34. Moody, J.R., Lindstrom, R.M.: ibid. *49*, 2264 (1977)
35. Laxen, D.P.H., Harrison, R.M.: ibid. *53*, 345 (1981)
36. Kuroha, T.: Bunseki Kagaku *21*, 502, 506 (1972)
37. Kuehner, E.C. et al.: Anal. Chem. *44*, 2050 (1972)
38. Mattinson, J.M.: ibid. *44*, 1715 (1972)
39. Little, K., Brooks, J.D.: ibid. *46*, 1343 (1974)
40. Dabeka, R.W. et al.: ibid. *48*, 1203 (1976)
41. Conway, B.E. et al.: ibid. *45*, 1331 (1973)
42. Irving, H., Cox, J.J.: Analyst *83*, 526 (1958)
43. Kwestroo, W., Visser, J.: ibid. *90*, 297 (1965)
44. Veillon, C., Reamer, D.C.: Anal. Chem. *53*, 549 (1981)
45. Hiraide, M., Mizuike, A.: Bull. Chem. Soc. Jpn. *48*, 3753 (1975)
46. Reiner, D., Poe, D.P.: Anal. Chem. *49*, 889 (1977)
47. LaFleur, P.D. (ed.): Accuracy in Trace Analysis: Sampling, Sample Handling, Analysis, 2 Vols., Nat. Bur. Stds. Special Publication 422 (1976)
48. Conway, E.J.: Microdiffusion Analysis and Volumetric Error, London, Crosby Lockwood 1950[3]
49. Werner, W., Tölg, G.: Fresenius Z. Anal. Chem. *276*, 103 (1975)
50. Kirkbright, G.F., Sargent, M.: Atomic Absorption and Fluorescence Spectroscopy, p. 637, London-New York-San Francisco, Academic Press 1974
51. Bricker, J.L.: Anal. Chem. *52*, 492 (1980)
52. Larsen, R.P., Ingber, N.M.: ibid. *31*, 1084 (1959)
53. Melnick, L.M. et al. (ed.): Determination of Gaseous Elements in Metals, New York, Wiley 1974
54. Grallath, E., Tölg, G.: Mikrochim. Acta *1978 II*, 547 (1978)
55. Luke, C.L.: Anal. Chem. *21*, 1369 (1949); *29*, 1227 (1957)
56. Mizuike, A., Kondo, A.: Mikrochim. Acta *1971*, 841 (1971)
57. Miwa, T. et al.: Anal. Chim. Acta *60*, 475 (1972)
58. Miwa, T. et al.: Bunseki Kagaku *27*, 228 (1978)
59. Terada, K. et al.: Bull. Chem. Soc. Jpn. *50*, 396 (1977)
60. Sakamoto, T. et al.: Bunseki Kagaku *24*, 457 (1975)
61. Miyazaki, A. et al.: Anal. Chim. Acta *90*, 119 (1977)
62. Thompson, M. et al.: Analyst *103*, 568, 705 (1978)
63. Pahlavanpour, B. et al.: ibid. *105*, 756 (1980); *106*, 467 (1981)
64. Andreae, M.O., Froelich, Jr., P.N.: Anal. Chem. *53*, 287 (1981)
65. Berndt, H. et al.: Fresenius Z. Anal. Chem. *296*, 377 (1979)
66. Drinkwater, J.E.: Analyst *101*, 672 (1976)
67. Jin, K. et al.: Bull. Chem. Soc. Jpn. *52*, 2276 (1979); *54*, 2934 (1981)
68. Subramanian, K.S.: Fresenius Z. Anal. Chem. *305*, 382 (1981)
69. Grallath, E. et al.: ibid. *302*, 40 (1980)
70. Holt, B.D.: Anal. Chem. *32*, 124 (1960)
71. Ishihara, Y. et al.: Bunseki Kagaku *11*, 566 (1962)
72. Ishihara, Y., Komuro, H.: ibid. *12*, 380 (1963)
73. Terada, K. et al.: Talanta *22*, 41 (1975)
74. Luke, C.L.: Anal. Chem. *30*, 1405 (1958)
75. Eberle, A.R., Lerner, M.W.: ibid. *32*, 146 (1960)
76. Freegarde, M., Cartwright, J.: Analyst *87*, 214 (1962)
77. Oldfield, J.H., Bridge, E.P.: ibid. *85*, 97 (1960)
78. Mykytiuk, A. et al.: Anal. Chem. *48*, 1462 (1976)
79. Hoffmeister, W.: Fresenius Z. Anal. Chem. *290*, 289 (1978)

80. Boutron, C.: Anal. Chim. Acta *61*, 140 (1972)
81. Boutron, C., Martin, S.: Anal. Chem. *51*, 140 (1979)
82. Mizuike, A., Takata, Y.: Bunseki Kagaku *12*, 1192 (1963)
83. Veleker, T.J., Mehalchick, E.J.: Anal. Chem. *33*, 767 (1961)
84. Miyamoto, M.: Bunseki Kagaku *12*, 233 (1963)
85. Kawasaki, K., Higo, M.: Anal. Chim. Acta *33*, 497 (1965)
86. Kometani, T.Y.: Anal. Chem. *49*, 2289 (1977)
87. Rees, W.T.: Analyst *87*, 202 (1962)
88. Pohl, F.A., Bonsels, W.: Mikrochim. Acta *1960*, 641 (1960); *1962*, 97 (1962)
89. Ferri, D., Buldini, P.L.: Anal. Chim. Acta *126*, 247 (1981)
90. Alimarin, I.P. (ed.): Analysis of High-Purity Materials, Jerusalem, Israel Program for Scientific Translations 1968
91. Luke, C.L., Campbell, M.E.: Anal. Chem. *25*, 1588 (1953); *28*, 1340 (1956)
92. Brophy, V.A. et al.: ibid. *26*, 430 (1954)
93. Heffelfinger, R.E. et al.: ibid. *34*, 621 (1962)
94. Oldfield, J.H., Mack, D.L.: Analyst *87*, 778 (1962)
95. Mack, D.L.: ibid. *88*, 481 (1963)
96. Itsuki, K., Kaji, T.: Bunseki Kagaku *8*, 703 (1959)
97. Bradshaw, G., Rands, J.: Analyst *85*, 76 (1960)
98. Nishimura, K. et al.: Bunseki Kagaku *13*, 220 (1964)
99. Bush, E.L.: Analyst *88*, 614 (1963)
100. Yokosuka, S., Shirakawa, S.: Bunseki Kagaku *7*, 363 (1958)
101. Schreiber, E.: Fresenius Z. Anal. Chem. *210*, 93 (1965)
102. Jones, R.H.: Analyst *71*, 60 (1946)
103. Itsuki, K. et al.: Bunseki Kagaku *9*, 840 (1960)
104. Miyamoto, M.: ibid. *10*, 98, 102 (1961)
105. Etten, N., Muschaweck, J.: Fresenius Z. Anal. Chem. *206*, 17 (1964)
106. Joshi, B.D. et al.: ibid. *260*, 107 (1972)
107. Bangia, T.R., Joshi, B.D.: ibid. *283*, 191 (1977)
108. Neeb, K.H.: ibid. *221*, 200 (1966)
109. Geilmann, W., Neeb, R.: Angew. Chem. *67*, 26 (1955)
110. Heinrichs, H.: Fresenius Z. Anal. Chem. *294*, 345 (1979)
111. Zil'bershtein, Kh.I. (ed.): Spectrochemical Analysis of Pure Substances, p. 245, Bristol, Adam Hilger Ltd. 1977
112. Grallath, E.: Fresenius Z. Anal. Chem. *300*, 97 (1980)
113. Winterkorn, M. et al.: Mikrochim. Acta, Suppl. 7, 27 (1977)
114. Pugh, H., Waterman, W.R.: Anal. Chim. Acta *55*, 97 (1971)
115. Harrison, T.S., Spikings, R.J.: ibid. *67*, 145 (1973)
116. Schoch, P. et al.: Fresenius Z. Anal. Chem. *271*, 12 (1974)
117. Ducret, L., Cornet, C.: Anal. Chim. Acta *25*, 542 (1961)
118. Meyer, A. et al.: Fresenius Z. Anal. Chem. *290*, 292 (1978); *305*, 1 (1981)
119. Han, H.B. et al.: Anal. Chim. Acta *128*, 9 (1981)
120. Han, H.B. et al.: ibid. *134*, 3 (1982)
121. Farzaneh, A. et al.: Fresenius Z. Anal. Chem. *296*, 383 (1979)
122. Geilmann, W. et al.: ibid. *154*, 418 (1957)
123. Geilmann, W.: ibid. *160*, 410 (1958)
124. Geilmann, W., Neeb, K.H.: ibid. *165*, 251 (1959)
125. Geilmann, W., Estebaranz, A.A.: ibid. *190*, 60 (1962)
126. Neeb, K.H.: ibid. *194*, 255 (1963)
127. Geilmann, W., Hepp, H.: ibid. *200*, 241 (1964)
128. Marshall, R.R., Hess, D.C.: Anal. Chem. *32*, 960 (1960)
129. Bächmann, K. et al.: Fresenius Z. Anal. Chem. *294*, 337 (1979); *296*, 374 (1979)
130. Keck, P.H. et al.: Anal. Chem. *28*, 995 (1956)
131. Mandelstam, S.: Appl. Spectry. *11*, 157 (1957)
132. O'Connell, R.F.: ibid. *18*, 179 (1964)

133. Harrison, S.H. et al.: Anal. Chem. *47*, 1685 (1975)
134. Hall, A., Godinho, M.C.: Anal. Chim. Acta *113*, 369 (1980)
135. Beyer, K.W., Aepli, O.T.: Anal. Chem. *29*, 1779 (1957)
136. Bächmann, K. et al.: Fresenius Z. Anal. Chem. *301*, 3 (1980)
137. Spachidis, C. et al.: ibid. *306*, 268 (1981)
138. Sinclair, V.M. et al.: Analyst *91*, 582 (1966)
139. Rajić, S.R., Marković, S.V.: Anal. Chim. Acta *50*, 169 (1970)
140. Takeuchi, T. et al.: J. Chem. Soc. Jpn., Ind. Chem. Sect. *64*, 1367 (1961)
141. Fukasawa, T. et al.: Bunseki Kagaku *22*, 280 (1973); *26*, 200 (1977)
142. Joshi, B.D. et al.: Mikrochim. Acta *1974*, 829 (1974)
143. Neeb, K.H.: Fresenius Z. Anal. Chem. *200*, 278 (1964)
144. Neeb, K.H.: ibid. *211*, 334 (1965)
145. Hirano, S. et al.: Bunseki Kagaku *18*, 574, 1270 (1969)
146. Koch, W.: Metallkundliche Analyse, Düsseldorf, Verlag Stahleisen 1965
147. Bächmann, K.: Talanta *29*, 1 (1982)
148. Morrison, G.H., Freiser, H.: Solvent Extraction in Analytical Chemistry, New York, Wiley 1957
149. Starý, J.: The Solvent Extraction of Metal Chelates, New York, Macmillan 1964
150. Marcus, Y., Kertes, A.S.: Ion Exchange and Solvent Extraction of Metal Complexes, London-New York-Sydney-Toronto, Wiley-Interscience 1969
151. Zolotov, Yu.A.: Extraction of Chelate Compounds, Ann Arbor, Humphery Science Publ. 1970
152. Koch, O.G., Koch-Dedic, G.A.: Handbuch der Spurenanalyse, Teil 1, Berlin, Springer 1974²
153. Sekine, T., Hasegawa, Y.: Solvent Extraction Chemistry Fundamentals and Applications, New York, Marcel Dekker 1977
154. Tanimura, T. et al.: Science *169*, 54 (1970)
155. Motojima, K.: Progress in Nuclear Energy, Ser. IX, Analytical Chemistry Vol. 8, Part 1, p. 49, Oxford, Pergamon 1967
156. Ringbom, A.: Complexation in Analytical Chemistry, New York, Wiley-Interscience 1963
157. Perrin, D.D.: Masking and Demasking of Chemical Reactions, New York, Wiley-Interscience 1970
158. Wyttenbach, A., Bajo, S.: Anal. Chem. *47*, 1813 (1975)
159. Klinkhammer, G.P.: ibid. *52*, 117 (1980)
160. Kanke, M. et al.: J. Chem. Soc. Jpn., Pure Chem. Sect. *92*, 983 (1971)
161. Ármannsson, H.: Anal. Chim. Acta *110*, 21 (1979)
162. Smith, Jr., R.G., Windom, H.L.: ibid. *113*, 39 (1980)
163. Shigematsu, T. et al.: Bunseki Kagaku *22*, 1162 (1973)
164. Korkisch, J., Sorio, A.: Anal. Chim. Acta *79*, 207 (1975)
165. Tweeten, T.N., Knoeck, J.W.: Anal. Chem. *48*, 64 (1976)
166. Lo, J.M., et al.: ibid. *49*, 1146 (1977)
167. Yamazaki, M. et al.: Bunseki Kagaku *30*, 40 (1981)
168. Gilbert, T.R., Clay, A.M.: Anal. Chim. Acta *67*, 289 (1973)
169. Aldous, K.M. et al.: Anal. Chem. *47*, 1034 (1975)
170. Jan, T.K., Young, D.R.: ibid. *50*, 1250 (1978)
171. Sperling, K.R.: Fresenius Z. Anal. Chem. *292*, 113 (1978)
172. Bone, K.M., Hibbert, W.D.: Anal. Chim. Acta *107*, 219 (1979)
173. Danielsson, L.-G. et al.: ibid. *98*, 47 (1978)
174. McLeod, C.W. et al.: Analyst *106*, 419 (1981)
175. Dornemann, A., Kleist, H.: Fresenius Z. Anal. Chem. *291*, 349 (1978)
176. Mizuno, T.: J. Chem. Soc. Jpn. *1973*, 1904 (1973)
177. Brooks, R.R.: Talanta *12*, 505, 511 (1965)
178. Jones, M. et al.: Anal. Chim. Acta *63*, 210 (1973)
179. Suzuki, M. et al.: Talanta *12*, 989 (1965)
180. Yanagisawa, M. et al.: ibid. *14*, 933 (1967)
181. Watanabe, K., Kawagaki, K.: Bull. Chem. Soc. Jpn. *48*, 1812 (1975)

182. Mereček, J., Singer, E.: Fresenius Z. Anal. Chem. *203*, 336 (1964)
183. Hubbard, G.L., Green, T.E.: Anal. Chem. *38*, 428 (1966)
184. Koch, O.G.: Mikrochim. Acta *1958*, 402 (1958)
185. Luke, C.L., Campbell, M.E.: Anal. Chem. *25*, 1588 (1953)
186. Payne, S.T.: Analyst *77*, 278 (1952)
187. Fowler, E.W.: ibid. *88*, 380 (1963)
188. Veleker, T.J.: Anal. Chem. *34*, 87 (1962)
189. Takahisa, M. et al.: Bunseki Kagaku *20*, 188 (1971)
190. Kawakubo, S. et al.: ibid. *30*, 594 (1981)
191. Fuller, C.W., Whitehead, J.: Anal. Chim. Acta *68*, 407 (1974)
192. Sanzolone, R.F. et al.: ibid. *105*, 247 (1979)
193. Koch, O.G.: Mikrochim. Acta *1958*, 92, 151, 347 (1958)
194. Apple, R.F., White, J.C.: Talanta *13*, 43 (1966)
195. Delves, H.T. et al.: Analyst *96*, 260 (1971)
196. Amore, F.: Anal. Chem. *46*, 1597 (1974)
197. Leitner, S.S., Savory, J.: Anal. Chim. Acta *74*, 133 (1975)
198. Willis, J.B.: Anal. Chem. *34*, 614 (1962)
199. Růžička, J., Starý, J.: Substoichiometry in Radiochemical Analysis, Oxford, Pergamon 1968
200. Donaldson, E.M., Inman, W.R.: Talanta *13*, 489 (1966)
201. Bankmann, E., Specker, H.: Fresenius Z. Anal. Chem. *162*, 18 (1958)
202. Carson, R.: Analyst *83*, 472 (1958)
203. Pohl, F.A., Bonsels, W.: Fresenius Z. Anal. Chem. *161*, 108 (1958)
204. Nishimura, K., Imai, T.: Bunseki Kagaku *13*, 518 (1964)
205. Marczenko, Z. et al.: Talanta *21*, 93 (1974)
206. Jackson, H., Phillips, D.S.: Analyst *87*, 712 (1962)
207. Parissakis, G., Issopoulos, P.B.: Mikrochim. Acta *1965*, 28 (1965)
208. Futekov, L. et al.: Fresenius Z. Anal. Chem. *306*, 381 (1981)
209. Hirano, S. et al.: Bunseki Kagaku *9*, 164 (1960)
210. Korenaga, T.: Mikrochim. Acta *1979 I*, 435 (1979)
211. Burke, K.E.: Analyst *97*, 19 (1972)
212. Burke, K.E.: Talanta *21*, 417 (1974)
213. Jackwerth, E.: Fresenius Z. Anal. Chem. *206*, 335 (1964)
214. Jackwerth, E., Schneider, E.-L.: ibid. *207*, 188 (1965)
215. Nakamura, H. et al.: Bunseki Kagaku *13*, 509 (1964)
216. Hirano, S. et al.: J. Chem. Soc. Jpn., Ind. Chem. Sect. *62*, 622 (1959)
217. Hirano, S. et al.: Bunseki Kagaku *11*, 1127 (1962)
218. Knapp, J.R. et al.: Anal. Chem. *34*, 1374 (1962)
219. Kuroha, T. et al.: Bunseki Kagaku *20*, 1137 (1971)
220. Tsukahara, I., Yamamoto, T.: Anal. Chim. Acta *61*, 33 (1972); *63*, 464 (1973); *135*, 235 (1982)
221. Tsukahara, I., Tanaka, M.: ibid. *116*, 383 (1980)
222. Tsukahara, I., Tanaka, M.: Talanta *27*, 655 (1980)
223. Tsukahara, I., Yamamoto, T.: ibid. *28*, 585 (1981)
224. Yang, M.H. et al.: J. Radioanal. Chem. *37*, 801 (1977)
225. Utsumi, S. et al.: J. Chem. Soc. Jpn. *1974*, 1073 (1974)
226. Heffelfinger, R.E. et al.: Anal. Chem. *30*, 112 (1958)
227. Rooney, R.C.: Analyst *83*, 83 (1958)
228. Répás, P. et al.: Fresenius Z. Anal. Chem. *207*, 263 (1965)
229. Owens, E.B.: Appl. Spectry. *13*, 105 (1959)
230. Oldfield, J.H., Bridge, E.P.: Analyst *86*, 267 (1961)
231. Monnier, D., Prod'hom, G.: Anal. Chim. Acta *31*, 101 (1964)
232. Ramamurty, C.K. et al.: Mikrochim. Acta *1980 I*, 79 (1980)
233. Kakumoto, S.: Bunseki Kagaku *13*, 1016 (1964)
234. Itsuki, K., Nishino, K.: ibid. *9*, 557 (1960)

235. Miyamoto, M.: ibid. *9*, 748, 753, 869 (1960); *10*, 217 (1961)
236. Cordis, V. et al.: Fresenius Z. Anal. Chem. *279*, 355 (1976)
237. Pohl, F.A., Bonsels, W.: Mikrochim. Acta *1961*, 314 (1961)
238. Miyamoto, M.: Bunseki Kagaku *9*, 925 (1960)
239. Duke, J.F.: Ultrapurification of Semiconductor Materials (ed.) Brooks, M.S., Kennedy, J.K., p. 356, New York, Macmillan 1962
240. Nishimura, K., Imai, T.: Bunseki Kagaku *13*, 423 (1964)
241. Jackwerth, E.: Fresenius Z. Anal. Chem. *202*, 81; *206*, 269 (1964)
242. Jackwerth, E.: ibid. *211*, 254 (1965)
243. Jackwerth, E.: ibid. *216*, 73 (1966)
244. Motojima, K. et al.: Anal. Chem. *34*, 571 (1962)
245. Osumi, Y. et al.: Bunseki Kagaku *20*, 1393 (1971)
246. Bangia, T.R. et al.: Fresenius Z. Anal. Chem. *310*, 410 (1982)
247. Startseva, E.A. et al.: ibid. *300*, 28 (1980)
248. Busev, A.I. et al.: Talanta *19*, 173 (1972)
249. Murata, K. et al.: Anal. Chem. *44*, 805 (1972)
250. Kawamoto, H., Akaiwa, H.: Chem. Lett. *1973*, 259 (1973)
251. Fujinaga, T. et al.: Bunseki Kagaku *18*, 398, 1113 (1969); *20*, 1255 (1971); *25*, 313 (1976)
252. Fujinaga, T. et al.: Bull. Chem. Soc. Jpn. *46*, 2090 (1973); *48*, 899 (1975)
253. Fujinaga, T., Puri, B.K.: Fresenius Z. Anal. Chem. *269*, 340 (1974)
254. Puri, B.K. et al.: Bull. Chem. Soc. Jpn. *52*, 3415 (1979)
255. Kawase, A. et al.: Bunseki Kagaku *30*, 229 (1981)
256. Barney II, J.E.: Anal. Chem. *27*, 1283 (1955)
257. Barney II, J.E., Haight, Jr., G.P.: ibid. *27*, 1285 (1955)
258. Nelson, K.H., Grimes, M.D.: ibid. *32*, 594 (1960)
259. Samuel, B.W., Brunnock, J.V.: ibid. *33*, 203 (1961)
260. Lancaster, W.A., Everingham, M.R.: ibid. *36*, 246 (1964)
261. Haas, C.S. et al.: ibid. *36*, 245 (1964)
262. Taguchi, I.: Bunseki *1979*, 370 (1979)
263. Kamada, H. (ed.): Modern Methods of State Analysis of Steels, Tokyo, AGNE Pub. Inc. 1979
264. Tsukahara, I. et al.: Bunseki Kagaku *18*, 1229, 1236 (1969)
265. Tsukahara, I.: ibid. *19*, 1496, 1502 (1970); *20*, 596 (1971); *21*, 370 (1972)
266. Tsukahara, I., Yabuki, E.: J. Japan Inst. Metals *36*, 66 (1972)
267. Analytical Methods Committee: Methods for the Determination of Trace Impurities in Aluminum, p. 152, Tokyo, Japan Light Metal Association, 1978
268. Matsumoto, K. et al.: Anal. Chim. Acta *115*, 149 (1980)
269. Simensen, C.J., Strand, G.: Fresenius Z. Anal. Chem. *308*, 11 (1981)
270. Matsumoto, K. et al.: Bunseki Kagaku *28*, 20 (1979)
271. Matsumoto, K. et al.: Fresenius Z. Anal. Chem. *309*, 398 (1981)
272. Matsumoto, K.: Anal. Chim. Acta *123*, 297 (1981)
273. Jackwerth, E., Kulok, A.: Fresenius Z. Anal. Chem. *257*, 28 (1971)
274. Jackwerth, E., Messerschmidt, J.: ibid. *274*, 205 (1975)
275. Jackwerth, E., Messerschmidt, J.: Anal. Chim. Acta *87*, 341 (1976)
276. Höhn, R., Jackwerth, E.: Erzmetall *29*, 279 (1976)
277. Jackwerth, E. et al.: Fresenius Z. Anal. Chem. *260*, 177 (1972)
278. Jackwerth, E. et al.: ibid. *264*, 1 (1973)
279. Höhn, R., Jackwerth, E.: ibid. *282*, 21 (1976)
280. Höhn, R. et al.: Spectrochim. Acta *29B*, 225 (1974)
281. Jackwerth, E. et al.: Anal. Chim. Acta *94*, 225 (1977)
282. Höhn, R., Jackwerth, E.: Fresenius Z. Anal. Chem. *289*, 47 (1978)
283. Matsumoto, K., Kiba, T.: Bunseki Kagaku *30*, 12 (1981)
284. Kiba, T. et al.: ibid. *24*, 116 (1975)
285. Matsumoto, K.: Fresenius Z. Anal. Chem. *305*, 370 (1981)
286. Luke, C.L., Flaschen, S.S.: Anal. Chem. *30*, 1406 (1958)

287. Parker, C.A., Barnes, W.J.: Analyst *85*, 828 (1960)
288. Morrison, G.H., Rupp, R.L.: Anal. Chem. *29*, 892 (1957)
289. Newton, D.C. et al.: Analyst *85*, 870 (1960)
290. Fukuda, K. et al.: Bunseki Kagaku *23*, 75 (1974)
291. Rynasiewicz, J. et al.: Anal. Chem. *26*, 935 (1954)
292. Mizuike, A. et al.: Talanta *19*, 527 (1972)
293. Mizuike, A., Fukuda, K.: Mikrochim. Acta *1975 I*, 281 (1975)
294. Mizuike, A., Kono, T.: ibid. *1970*, 665 (1970)
295. Mizuike, A. et al.: ibid. *1970*, 1095 (1970)
296. Mizuike, A., Fukuda, K.: ibid. *1972*, 257 (1972)
297. Furman, N.H. (ed.): Standard Methods of Chemical Analysis, Vol. I, p. 475, Princeton, N.J.-New York-Toronto-London, D. Van Nostrand 1962[6]
298. Beamish, F.E.. Talanta *14*, 1133 (1967)
299. Beamish, F.E., Van Loon, J.C.: Recent Advances in the Analytical Chemistry of the Noble Metals, Oxford, Pergamon 1972
300. Davies, B.E. (ed.): Applied Soil Trace Elements, Chichester-New York-Brisbane-Toronto, Wiley 1980
301. Salutsky, M.L.: Precipitates – Their Formation, Properties, and Purity, in: Treatise on Analytical Chemistry (ed.) Kolthoff, I.M. and Elving, P.J., Part I, Vol. 1, Chap. 18, p. 733, New York, Interscience 1959
302. Hermann, J.A., Suttle, J.F.: Precipitation and Crystallization, in: Treatise on Analytical Chemistry (ed.) Kolthoff, I.M. and Elving, P.J., Part I, Vol. 3, Chap. 32, p. 1367, New York, Interscience 1961
303. Kolthoff, I.M. et al.: Quantitative Chemical Analysis, London, Macmillan 1969[4]
304. Gordon, L. et al.: Precipitation from Homogeneous Solution, New York, Wiley 1959
305. Jackwerth, E.: Pure & Appl. Chem. *51*, 1149 (1979)
306. Englis, D.T., Burnett, B.B.: Anal. Chim. Acta *13*, 574 (1955)
307. Katsura, T.: Bunseki Kagaku *10*, 1323 (1961)
308. Jackwerth, E. et al.: Fresenius Z. Anal. Chem. *260*, 101 (1972)
309. Meyer, J.: ibid. *219*, 147 (1966)
310. Meyer, J.: ibid. *231*, 241 (1967)
311. Luke, C.L.: Anal. Chem. *32*, 836 (1960)
312. Matano, N., Kawase, A.: Bunseki Kagaku *11*, 346 (1962)
313. Kawase, A.: ibid. *11*, 1162 (1962)
314. Luke, C.L.: Anal. Chem. *27*, 1150 (1955)
315. Hair, R.P., Newman, E.J.: Analyst *89*, 42 (1964)
316. Jackwerth, E., Willmer, P.G.: Fresenius Z. Anal. Chem. *279*, 23 (1976)
317. Shigetomi, Y., Hirota, Y.: Bunseki Kagaku *24*, 148 (1975)
318. Marczenko, Z.: Crit. Rev. Anal. Chem. *10*, 195 (1981)
319. Luke, C.L.: Anal. Chim. Acta *41*, 237 (1968)
320. Hirokawa, K.: Fresenius Z. Anal. Chem. *260*, 4 (1972)
321. Kessler, J.E., Mitchell, J.W.: Anal. Chem. *50*, 1644 (1978)
322. Kessler, J.E. et al.: Talanta *26*, 21 (1979)
323. Ujihira, Y.: J. Chem. Soc. Jpn., Pure Chem. Sect. *84*, 642 (1963)
324. Okochi, H., Sudo, E.: Bunseki Kagaku *22*, 431 (1973)
325. Chan, Y.K., Riley, J.P.: Anal. Chim. Acta *33*, 36 (1965)
326. Chan, K.M., Riley, J.P.: ibid. *34*, 337 (1966)
327. Chuecas, L., Riley, J.P.: ibid. *35*, 240 (1966)
328. Kouimtzis, Th.A. et al.: ibid. *123*, 315 (1981)
329. Shimizu, T. et al.: Bunseki Kagaku *30*, 113 (1981)
330. Hudnik, V. et al.: Anal. Chim. Acta *98*, 39 (1978)
331. Bruninx, E., Van Meyl, E.: ibid. *80*, 85 (1975)
332. Kaneko, E.: Bunseki Kagaku *27*, 250 (1978)
333. Shimizu, T., Sakai, K.: J. Chem. Soc. Jpn. *1981*, 26 (1981)
334. Shigematsu, T. et al.: ibid. Pure Chem. Sect. *85*, 490 (1964)

335. Sato, A., Saitoh, N.: Bunseki Kagaku 25, 663 (1976)
336. Abu-Hilal, A.H., Riley, J.P.: Anal. Chim. Acta 131, 175 (1981)
337. Tsuyama, A., Nakashima, S.: Bunseki Kagaku 29, 81 (1980)
338. Shigetomi, Y.: ibid. 24, 699 (1975)
339. Portmann, J.E., Riley, J.P.: Anal. Chim. Acta 35, 35 (1966)
340. Chan, K.M., Riley, J.P., ibid. 36, 220 (1966)
341. Ishibashi, M. et al.: J. Chem. Soc. Jpn., Pure Chem. Sect. 83, 295 (1962)
342. Owa, T. et al.: Bunseki Kagaku 21, 878 (1972)
343. Reinhardt, K.H., Müller, H.J.: Fresenius Z. Anal. Chem. 292, 359 (1978)
344. Perry, D.L. et al.: Anal. Chem. 53, 1048 (1981)
345. Shigematsu, T. et al.: J. Chem. Soc. Jpn., Pure Chem. Sect. 84, 336 (1963)
346. Weiss, H.V., Lai, M.G.: Anal. Chim. Acta 28, 242 (1963)
347. Weiss, H.V. et al.: ibid. 25, 550 (1961)
348. Weiss, H.V., Lai, M.G.: Talanta 8, 72 (1961)
349. Lai, M.G., Weiss, H.V.: Anal. Chem. 34, 1012 (1962)
350. Portmann, J.E., Riley, J.P.: Anal. Chim. Acta 31, 509 (1964)
351. Gohda, S.: Bull. Chem. Soc. Jpn. 45, 1704 (1972)
352. Liardon, O., Ryan, D.E.: Anal. Chim. Acta 83, 421 (1976)
353. Weiss, H.V. et al.: ibid. 104, 337 (1979)
354. Shigematsu, T. et al.: Bull. Chem. Soc. Jpn. 41, 609 (1968)
355. Riley, J.P., Topping, G.: Anal. Chim. Acta 44, 234 (1969)
356. Silvey, W.D., Brennan, R.: Anal. Chem. 34, 784 (1962)
357. Vanderstappen, M.G., Van Grieken, R.E.: Talanta 25, 653 (1978)
358. Watanabe, H. et al.: ibid. 19, 1363 (1972)
359. Panayappan, R. et al.: Anal. Chem. 50, 1125 (1978)
360. Watanabe, H., Ueda, T.: Bull. Chem. Soc. Jpn. 53, 411 (1980)
361. Yang, C.Y. et al.: Analyst 106, 385 (1981)
362. Takemoto, S. et al.: Bunseki Kagaku 25, 40 (1976)
363. Scheubeck, E.: Mikrochim. Acta 1980 II, 283 (1980)
364. Scheubeck, E. et al.: Fresenius Z. Anal. Chem. 303, 257 (1980)
365. Pik, A.J. et al.: Anal. Chim. Acta 110, 61 (1979)
366. Boyle, E.A., Edmond, J.M.: ibid. 91, 189 (1977)
367. Krishnamurty, K.V., Reddy, M.M.: Anal. Chem. 49, 222 (1977)
368. Caravajal, G.S. et al.: Anal. Chim. Acta 135, 205 (1982)
369. Satake, M. et al.: Mikrochim. Acta 1980 I, 455 (1980)
370. Tanaka, K.: Bunseki Kagaku 10, 612 (1961)
371. Itsuki, K. et al.: ibid. 8, 804 (1959); 9, 840 (1960)
372. Sudo, N., Inoue, S.: ibid. 18, 717 (1969)
373. Okochi, H., Sudo, E.: ibid. 20, 683 (1971)
374. Yanagihara, T. et al.: ibid. 10, 467 (1961)
375. Ishibashi, M. et al.: ibid. 7, 553 (1958)
376. Okochi, H., Sudo, E.: ibid. 18, 1376 (1969); 19, 659 (1970)
377. Kato, K.: ibid. 30, 73 (1981)
378. Ko, R., Anderson, P.: Anal. Chem. 41, 177 (1969)
379. Reichel, W., Bleakley, B.G.: ibid. 46, 59 (1974)
380. Sudo, E., Okochi, H.. Bunseki Kagaku 18 , 501 (1969)
381. Sudo, E., Ogawa, H.: J. Japan Inst. Metals 28, 421 (1964)
382. Young, P.N.W.. Analyst 99, 588 (1974)
383. Yoshimura, W.: Bunseki Kagaku 30, 347 (1981)
384. Mizuike, A. et al.: Mikrochim. Acta 1974, 915 (1974)
385. Wurzinger, H., Müller, K.: Fresenius Z. Anal. Chem. 284, 101 (1977)
386. Harada, Y.: Bunseki Kagaku 31, 130 (1982)
387. Koch, O.G.: Fresenius Z. Anal. Chem. 265, 29 (1973)
388. Miyamoto, M.: Bunseki Kagaku 10, 438 (1961)
389. Luke, C.L.: Anal. Chem. 31, 1680 (1959)

390. Blakeley, St.J.H. et al.: ibid. *45*, 1941 (1973)
391. Hirose, A., Ishii, D.: J. Radioanal. Chem. *20*, 17 (1974)
392. Ko. R., Weiler, M.R.: Anal. Chem. *34*, 85 (1962)
393. Kallmann, S. et al.: ibid. *32*, 1278 (1960)
394. Nakajima, T. et al.: Bunseki Kagaku *19*, 1183 (1970)
395. Ishii, D., Takeuchi, T.: ibid. *11*, 174 (1962)
396. Yanagihara, T. et al.: ibid. *8*, 576 (1959)
397. Jackwerth, E.: Fresenius Z. Anal. Chem. *235*, 235 (1968)
398. Jackwerth, E. et al.: ibid. *247*, 149 (1969)
399. Jackwerth, E.: ibid. *251*, 353 (1970)
400. Luke, C.L.: Anal. Chem. *31*, 572 (1959)
401. Grünwald, P. et al.: Fresenius Z. Anal. Chem. *279*, 187 (1976)
402. Kujirai, O. et al.: Talanta *29*, 27 (1982)
403. Hirano, S., Ujihira, Y.: J. Chem. Soc. Jpn., Ind. Chem. Sect. *66*, 939 (1963)
404. Veale, C.R., Wood, R.G.: Analyst *85*, 371 (1960)
405. Jackwerth, E. et al.: Fresenius Z. Anal. Chem. *299*, 362 (1979)
406. Pakalns, P.: Anal. Chim. Acta *57*, 51 (1971)
407. Yoshikawa, S., Nakamura, R.: Bunseki Kagaku *30*, 17 (1981)
408. Berndt, H., Messerschmidt, J.: Fresenius Z. Anal. Chem. *299*, 28 (1979)
409. Dehm, R.L. et al.: Anal. Chem. *33*, 607 (1961)
410. Farquhar, M.C. et al.: ibid. *38*, 208 (1966)
411. Jackwerth, E., Salewski, S.: Fresenius Z. Anal. Chem. *310*, 108 (1982)
412. Volland, G. et al.: Anal. Chim. Acta *90*, 15 (1977)
413. Batley, G.E., Matousek, J.P.: Anal. Chem. *49*, 2031 (1977); *52*, 1570 (1980)
414. Torsi, G. et al.: Anal. Chim. Acta *124*, 143 (1981)
415. Lingane, J.J.: Electroanalytical Chemistry, New York, Interscience 1958[2]
416. Neeb, R.: Inverse Polarographie und Voltammetrie, Weinheim/Bergstr., Verlag Chemie GmbH 1969
417. Brainina, Kh. Z.: Stripping Voltammetry in Chemical Analysis, New York, Wiley 1974
418. Vydra, F. et al.: Electrochemical Stripping Analysis, Chichester, Ellis Horwood 1976
419. Mizuike, A. et al.: Bull. Chem. Soc. Jpn. *42*, 253 (1969)
420. Thomassen, Y. et al.: Anal. Chim. Acta *83*, 103 (1976)
421. Mizuike, A. et al.: Radioisotopes [Tokyo] *17*, 199 (1968)
422. Vassos, B.H. et al.: Anal. Chem. *45*, 792 (1973)
423. Boslett, Jr., J.A. et al.: ibid. *49*, 1734 (1977)
424. Lux, F.: Radiochim. Acta *1*, 20 (1962)
425. Lund, W. et al.: Anal. Chim. Acta *81*, 319 (1976)
426. Malissa, H., Marr, I.L.: Mikrochim. Acta *1971*, 241 (1971)
427. Lund, W., Larsen, B.V.: Anal. Chim. Acta *70*, 299 (1974); *72*, 57 (1974)
428. Schaller, K.-H. et al.: Fresenius Z. Anal. Chem. *256*, 123 (1971)
429. Miwa, T. et al.: Bunseki Kagaku *27*, 228 (1978)
430. Mizuike, A., Mitsuya, N.: ibid. *17*, 1259 (1968)
431. Maxwell, J.A., Graham, R.P.: Chem. Rev. *46*, 471 (1950)
432. Page, J.A. et al.: Analyst *87*, 245 (1962)
433. Alimarin, I.P., Petrikova, M.N.: Inorganic Ultramicroanalysis, Oxford, Pergamon 1964
434. Herringshaw, J.F., Kassir, Z.M.: Analyst *87*, 923 (1962)
435. Schmidt, W.E., Bricker, C.E.: J. Electrochem. Soc. *102*, 623 (1955)
436. Erdey-Gruz, T., Vazsonyi-Zilahy, A.: Z. Physik. Chem. *A177*, 292 (1936)
437. Furman, N.H. et al.: J. Wash. Acad. Sci. *38*, 159 (1948)
438. Casto, C.C.: Analytical Chemistry of the Manhattan Project (ed.) Rodden, C.J., p. 511, New York, McGraw-Hill 1950
439. Sambucetti, C.J. et al.: Proc. Intern. Conf. Peaceful Uses At. Energy, Geneva 1955, Vol. 8, p. 266, New York, United Nations 1956
440. Mizuike, A., Hirano, S.: Bunseki Kagaku *7*, 545 (1958)
441. Mizuike, A., Hirano, S.: Proc. 2nd Symp. At. Energy, Tokyo, Vol. 3, p. 6, 1958

442. Hirano, S. et al.: J. Chem. Soc. Jpn., Ind. Chem. Sect. *62*, 1491 (1959)
443. Hirano, S. et al.: Bunseki Kagaku *8*, 827 (1959)
444. Malvano, R.: Anal. Chim. Acta *38*, 341 (1967)
445. Taylor, J.K., Smith, S.W.: J. Res. Natl. Bur. Standards *56*, 301 (1956)
446. Tölg, G.: Talanta *21*, 327 (1974)
447. Jensen, F.O. et al.: Anal. Chim. Acta *72*, 245 (1974)
448. Batley, G.E.: ibid. *124*, 121 (1981)
449. Jørstad, K., Salbu, B.: Anal. Chem. *52*, 672 (1980)
450. Center, E.J. et al.: ibid. *23*, 1134 (1951)
451. Spitz, E.W. et al.: ibid. *26*, 304 (1954)
452. Donaldson, E.M.: Talanta *28*, 461 (1981)
453. Chirnside, R.C. et al.: Analyst *82*, 18 (1957)
454. Rooney, R.C.: ibid. *83*, 546 (1958)
455. Meites, L.: Anal. Chem. *27*, 977 (1955)
456. Hirano, S., Mizuike, A.: Bunseki Kagaku *8*, 746 (1959)
457. Mizuike, A. et al.: Mikrochim. Acta *1973*, 291 (1973)
458. Hirano, S., Mizuike, A.: J. Chem. Soc. Jpn., Ind. Chem. Sect. *62*, 1497 (1959)
459. Mizuike, A.: Talanta *9*, 948 (1962)
460. Mizuike, A., Ujihira, Y.: Bunseki Kagaku *12*, 748 (1963)
461. Mizuike, A. et al.: Mikrochim. Acta *1971*, 783 (1971)
462. Mizuike, A. et al.: Bull. Chem. Soc. Jpn. *46*, 3596 (1973)
463. Gantchev, N., Dimitrova, A.: Mikrochim. Acta *1969*, 1257 (1969)
464. Fishman, M.J.: Anal. Chem. *42*, 1462 (1970)
465. Dogan, S., Haerdi, W.: Anal. Chim. Acta *76*, 345 (1975); *84*, 89 (1976)
466. Kurosawa, F. et al.: J. Japan Inst. Metals *43*, 1068 (1979); *44*, 539, 677 (1980)
467. Keulemans, A.I.M.: Gas Chromatography, p. 96, New York, Reinhold Pub. Corp. 1957
468. Guedes da Mota, M.M. et al.: Fresenius Z. Anal. Chem. *285*, 238 (1977); *287*, 19 (1977)
469. Samuelson, O.: Ion Exchange Separations in Analytical Chemistry, Stockholm, Almqvist & Wiksell and New York, Wiley 1963
470. Inczédy, J.: Analytical Applications of Ion Exchangers, Oxford, Pergamon 1966
471. Rieman III, W., Walton, H.F.: Ion Exchange in Analytical Chemistry, Oxford, Pergamon 1970
472. Braun, T., Farag, A.B.: Anal. Chim. Acta *99*, 1 (1978)
473. Moody, G.J., Thomas, J.D.R.: Analyst *104*, 1 (1979)
474. Leitch, R.E., DeStefano, J.J.: J. Chromatog. Sci. *11*, 105 (1973)
475. Schmuckler, G.: Talanta *12*, 281 (1965)
476. Hirsch, R.F. et al.: ibid. *17*, 483 (1970)
477. Fritz, J.S., Moyers, E.M.: ibid. *23*, 590 (1976)
478. Vernon, F., Eccles, H.: Anal. Chim. Acta *63*, 403 (1973)
479. Dingman, Jr., J.F. et al.: Anal. Chem. *46*, 774 (1974)
480. Barnes, R.M., Genna, J.S.: ibid. *51*, 1065 (1979)
481. Miyazaki, A., Barns, R.M.: ibid. *53*, 299, 364 (1981)
482. Ghosh, J.P., Das, H.R.: Talanta *28*, 274 (1981)
483. Ghosh, J.P. et al.: ibid. *28*, 957 (1981)
484. Colella, M.B. et al.: Anal. Chem. *52*, 967, 2347 (1980)
485. Koster, G., Schmuckler, G.: Anal. Chim. Acta *38*, 179 (1967)
486. Griesbach, M., Lieser, K.H.: Fresenius Z. Anal. Chem. *302*, 184 (1980)
487. Lieser, K.H., Thybusch, D.: ibid. *306*, 100 (1981)
488. Kraus, K.A., Nelson, F.: Proc. Intern. Conf. Peaceful Uses At. Energy, Geneva 1955, Vol. 7, p. 113, New York, United Nations 1956
489. Nelson, F. et al.: J. Chromatog. *13*, 503 (1964)
490. Ruch, R.R. et al.: Anal. Chem. *36*, 2311 (1964)
491. Strelow, F.W.E.: ibid. *32*, 1185 (1960)
492. Nelson, F., Michelson, D.C.: J. Chromatog. *25*, 414 (1966)
493. Strelow, F.W.E. et al.: Anal. Chem. *37*, 106 (1965)

494. Korkisch, J., Ahluwalia, S.S.: Talanta *14*, 155 (1967)
495. Korkisch, J., Klakl, E.: ibid. *16*, 377 (1969)
496. Korkisch, J. et al.: ibid. *14*, 1069 (1967)
497. Faris, J.P.: Anal. Chem. *32*, 520 (1960)
498. Faris, J.P., Buchanan, R.F.: ibid. *36*, 1157 (1964)
499. Strelow, F.W.E., Bothma, C.J.C.: ibid. *39*, 595 (1967)
500. Nelson, F. et al.: J. Am. Chem. Soc. *82*, 339 (1960)
501. Huff, E.A.: Anal. Chem. *36*, 1921 (1964)
502. Van Den Winkel, P. et al.: Anal. Chim. Acta *56*, 241 (1971)
503. Klakl, E., Korkisch, J.: Talanta *16*, 1177 (1969)
504. Akaiwa, H. et al.: Chem. Lett. *1975*, 1049 (1975)
505. Akaiwa, H. et al.: J. Radioanal. Chem. *36*, 59 (1977)
506. Akaiwa, H. et al.: Talanta *24*, 394 (1977)
507. Akaiwa, H. et al.: Radioisotopes [Tokyo] *28*, 291, 681 (1979); *29*, 521 (1980)
508. Going, J.E. et al.: Anal. Chim. Acta *81*, 349 (1976)
509. Berge, D.G., Going, J.E.: ibid. *123*, 19 (1981)
510. Tanaka, H. et al.: Talanta *23*, 489 (1976)
511. Chikuma, M. et al.: ibid. *27*, 807 (1980)
512. Lee, K.S. et al.: Anal. Chem. *50*, 255 (1978)
513. Gohda, S. et al.: Bunseki Kagaku *28*, 485 (1979)
514. Brajter, K., Dabek-Zlotorzyńska, E.: Talanta *27*, 19 (1980)
515. Small, H. et al.: Anal. Chem. *47*, 1801 (1975)
516. Lochmüller, C.H. et al.: ibid. *46*, 440 (1974)
517. Yoshimura, K. et al.: Talanta *23*, 449 (1976); *25*, 579 (1978); *27*, 693 (1980); *29*, 173 (1982)
518. Yoshimura, K., Ohashi, S.: ibid. *25*, 103 (1978)
519. Toshimitsu, Y. et al.: ibid. *26*, 273 (1979)
520. Nigo, S. et al.: ibid. *28*, 669 (1981)
521. Tuck, B., Osborn, E.M.: Analyst *85*, 105 (1960)
522. McNutt, N.S., Maier, R.H.: Anal. Chem. *34*, 276 (1962)
523. James, H.: Analyst *98*, 274 (1973)
524. Smales, A.A., Salmon, L.: ibid. *80*, 37 (1955)
525. Ramseyer, G.O., Janauer, G.E.: Anal. Chim. Acta *77*, 133 (1975)
526. Wickbold, R.: Fresenius Z. Anal. Chem. *171*, 81 (1959)
527. Westland, A.D., Langford, R.R.: Anal. Chem. *28*, 1996 (1956)
528. Brooks, R.R.: Analyst *85*, 745 (1960)
529. Portmann, J.E., Riley, J.P.: Anal. Chim. Acta *34*, 201 (1966)
530. Hiiro, K. et al.: Bunseki Kagaku *22*, 1210 (1973)
531. Kawabuchi, K., Riley, J.P.: Anal. Chim. Acta *65*, 271 (1973)
532. Becknell, D.E. et al.: Anal. Chem. *43*, 1230 (1971)
533. Sanemasa, I. et al.: Anal. Chim. Acta *130*, 149 (1981)
534. Korkisch, J., Sorio, A.: ibid. *76*, 393 (1975)
535. Kiriyama, T., Kuroda, R.: Fresenius Z. Anal. Chem. *288*, 354 (1977)
536. Kiriyama, T. et al.: ibid. *307*, 352 (1981)
537. Tanaka, T. et al.: Bunseki Kagaku *28*, 43 (1979); *30*, 131 (1981)
538. Kelso, F.S. et al.: Anal. Chem. *36*, 577 (1964)
539. Korkisch, J., Krivanec, H.: Talanta *23*, 295 (1976)
540. Abe, M. et al.: Bull Chem. Soc. Jpn. *51*, 1090 (1978)
541. Biechler, D.G.: Anal. Chem. *37*, 1054 (1965)
542. Riley, J.P., Taylor, D.: Anal. Chim. Acta *40*, 479 (1968)
543. Lee, C. et al.: Talanta *24*, 241 (1977)
544. Kingston, H.M. et al.: Anal. Chem. *50*, 2064 (1978)
545. Sturgeon, R.E. et al.: Talanta *27*, 85 (1980)
546. Sturgeon, R.E. et al.: Anal. Chem. *52*, 1585 (1980)
547. Berman, S.S. et al.: ibid. *52*, 488 (1980)

548. Pakalns, P.: Anal. Chim. Acta *120*, 289 (1980)
549. Rasmussen, L.: ibid. *125*, 117 (1981)
550. Hirose, A. et al.: J. Chem. Soc. Jpn. *1974*, 900 (1974)
551. Hirose, A. et al.: Anal. Chim. Acta *97*, 303 (1978)
552. Figura, P., McDuffie, B.: Anal. Chem. *49*, 1950 (1977)
553. Van Grieken, R.E. et al.: ibid. *49*, 1326 (1977)
554. Yamagami, E. et al.: Analyst *105*, 491 (1980)
555. Raynolds, G.F.: ibid. *82*, 46 (1957)
556. Iida, Y., Mizuike, A.: Bunseki Kagaku *13*, 68 (1964)
557. Van Niekerk, J.N. et al.: Anal. Chem. *33*, 213 (1961)
558. Buehwald, H., Wood, L.G.: ibid. *25*, 664 (1953)
559. Cranston, H.A., Thompson, J.B.: Ind. Eng. Chem., Anal. Ed. *18*, 323 (1946)
560. Bosholm, J.: Anal. Chim. Acta *34*, 71 (1966)
561. Hettel, H.J., Fassel, V.A.: Anal. Chem. *27*, 1311 (1955)
562. Grossmann, O. et al.: Fresenius Z. Anal. Chem. *219*, 48 (1966)
563. Spano, E.F., Green, T.E.: Anal. Chem. *38*, 1341 (1966)
564. Fukasawa, T., Yamane, T.: Bunseki Kagaku *24*, 120 (1975)
565. Kitazume, E. et al.: ibid. *27*, 566 (1978)
566. Okochi, H.: ibid. *20*, 1381 (1971)
567. Miwa, T. et al.: Talanta *17*, 108 (1970)
568. Mizuike, A. et al.: Proc. 2nd Symp. At. Energy, Tokyo, Vol. 3., p. 9, 1958
569. Mizuike, A. et al.: J. Chem. Soc. Jpn., Ind. Chem. Sect. *61*, 1459 (1958); *67*, 2042 (1964)
570. Mizuike, A. et al.: Anal. Chim. Acta *32*, 428 (1965)
571. Fukasawa, T. et al.: Bunseki Kagaku *17*, 713 (1968)
572. Hirose, A., Ishii, D.: J. Chem. Soc. Jpn. *1972*, 2364 (1972); *1974*, 2351 (1974)
573. Hirose, A., Ishii, D.: J. Radioanal. Chem. *41*, 37 (1977)
574. Miyamoto, M.: Bunseki Kagaku *10*, 321 (1961)
575. Yanagihara, T. et al.: ibid. *9*, 439, 539 (1960)
576. Kawase, A., Ogawa, H.: ibid. *11*, 1155 (1962)
577. Yoshino, Y., Kojima, M.: ibid. *6*, 160 (1957)
578. Fukasawa, T. et al.: ibid. *19*, 1417 (1970)
579. Fukasawa, T., Katagiri, K.: ibid. *21*, 480 (1972)
580. Imoto, H.: ibid. *10*, 124, 1354 (1961)
581. Pietri, C.E., Wenzel, A.W.: Anal. Chem. *35*, 209 (1963)
582. Wenzel, A.W., Pietri, C.E.: ibid. *36*, 2083 (1964)
583. Fukasawa, T. et al.: Bunseki Kagaku *20*, 193 (1971)
584. Birks, F.T. et al.: Analyst *89*, 36 (1964)
585. Nakashima, F.: Anal. Chim. Acta *30*, 167, 255 (1964)
586. Huff, E.A.: Anal. Chem. *37*, 533 (1965)
587. Mizuike, A. et al.: Bunseki Kagaku *21*, 1645 (1972)
588. Sakamoto, T. et al.: ibid. *19*, 1218 (1970)
589. Titze, H.: Mikrochim. Acta *1977 I*, 475 (1977)
590. Yoshino, Y., Kojima, M.: Bunseki Kagaku *4*, 311 (1955)
591. Chang, C.C. et al.: Fresenius Z. Anal. Chem. *270*, 187 (1974)
592. Tera, F. et al.: Anal. Chem. *37*, 358 (1965)
593. Ruch, R.R. et al.: ibid. *37*, 1565 (1965)
594. Brody, J.K. et al.: ibid. *30*, 1909 (1958)
595. Nakashima, F.: Anal. Chim. Acta *28*, 54 (1963)
596. Mizuike, A. et al.: ibid. *44*, 425 (1969)
597. Goode, G.C., Campbell, M.C.: ibid. *27*, 422 (1962)
598. Muzzarelli, R.A.A., Rocchetti, R.: ibid. *64*, 371 (1973)
599. Lieser, K.H. et al.: Fresenius Z. Anal. Chem. *284*, 361 (1977)
600. Burba, P., Lieser, K.H.: ibid. *286*, 191 (1977); *297*, 374 (1979); *298*, 373 (1979)
601. Burba, P. et al.: ibid. *289*, 28 (1978); *291*, 273 (1978)
602. Smits, J.A., Van Grieken, R.E.: Anal. Chem. *52*, 1479 (1980)

603. Smits, J., Van Grieken, R.: Anal. Chim. Acta *123*, 9 (1981)
604. Förster, M., Lieser, K.H.: Fresenius Z. Anal. Chem. *309*, 355 (1981)
605. Maloney, M.P. et al.: Analyst *105*, 1087 (1980)
606. Braun, T., Abbas, M.N.: Anal. Chim. Acta *119*, 113 (1980)
607. Mazurski, M.A.J. et al.: ibid. *65*, 99 (1973)
608. Yoshimura, K. et al.: ibid. *109*, 115 (1979); *130*, 345 (1981)
609. Mizuike, A., Fukuda, K.: ibid. *44*, 193 (1969)
610. Fukuda, K., Mizuike, A.: ibid. *51*, 77 (1970); *67*, 207 (1973)
611. Mizuike, A. et al.: Mikrochim. Acta *1979 II*, 487 (1979)
612. Fujinaga, T. et al.: Talanta *26*, 964 (1979)
613. Fukuda, K., Mizuike, A.: Bunseki Kagaku *17*, 65 (1968); *18*, 1130 (1969)
614. Fukuda, K. et al.: Radioisotopes [Tokyo] *19*, 247 (1970)
615. Cheng, K.L., Guh, H.Y.: Mikrochim. Acta *1978 I*, 55 (1978)
616. Terada, K. et al.: Bull. Chem. Soc. Jpn. *50*, 1060 (1977); *53*, 1605 (1980)
617. Terada, K. et al.: Anal. Chim. Acta *116*, 127 (1980)
618. Terada, K., Nakamura, K.: Talanta *28*, 123 (1981)
619. Leyden, D.E., Luttrell, G.H.: Anal. Chem. *47*, 1612 (1975)
620. Leyden, D.E. et al.: ibid. *48*, 67 (1976)
621. Leyden, D.E. et al.: Anal. Chim. Acta *84*, 97 (1976)
622. Guedes da Mota, M.M. et al.: Fresenius Z. Anal. Chem. *287*, 19 (1977); *296*, 345 (1979)
623. Jezorek, J.R., Freiser, H.: Anal. Chem. *51*, 366 (1979)
624. Hirayama, K., Unohara, N.: Bunseki Kagaku *29*, 452 (1980)
625. Sturgeon, R.E. et al.: Anal. Chem. *53*, 2337 (1981)
626. Topping, J.J., MacCrehan, W.A.: Talanta *21*, 1281 (1974)
627. Taguchi, S., Goto, K.: ibid. *27*, 819 (1980)
628. Taguchi, S. et al.: ibid. *28*, 613 (1981)
629. Watanabe, H. et al.: Anal. Chem. *53*, 738 (1981)
630. Sturgeon, R.E. et al.: Talanta *29*, 167 (1982)
631. Lorber, K., Müller, K.: Mikrochim. Acta *1976 I*, 375 (1976)
632. Krefeld, R. et al.: ibid. *1965*, 133 (1965)
633. Jackwerth, E. et al.: Fresenius Z. Anal. Chem. *266*, 1 (1973)
634. Koshima, H., Onishi, H.: Talanta *27*, 795 (1980)
635. Kimura, M., Kawanami, K.: J. Chem. Soc. Jpn. *1981*, 1 (1981)
636. Vanderborght, B.M., Van Grieken, R.E.: Anal. Chem. *49*, 311 (1977)
637. Jackwerth, E., Berndt, H.: Anal. Chim. Acta *74*, 299 (1975)
638. Berndt, H., Jackwerth, E.: Fresenius Z. Anal. Chem. *290*, 369 (1978)
639. Berndt, H. et al.: Anal. Chim. Acta *93*, 45 (1977)
640. Jackwerth, E.: Fresenius Z. Anal. Chem. *271*, 120 (1974)
641. Berndt, H., Messerschmidt, J.: ibid. *308*, 104 (1981)
642. Berndt, H. et al.: ibid. *310*, 230 (1982)
643. Amphlett, C.B.: Inorganic Ion Exchangers, Amsterdam, Elsevier 1964
644. Veselý, V., Pekárek, V.: Talanta *19*, 219, 1245 (1972)
645. Šulcek, Z., Sixta, V.: Anal. Chim. Acta *53*, 335 (1971)
646. Lieser, K.H. et al.: Fresenius Z. Anal. Chem. *298*, 378 (1979)
647. Fukuda, K., Mizuike, A.: Anal. Chim. Acta *51*, 527 (1970)
648. Miwa, T. et al.: Bunseki Kagaku *19*, 786 (1970)
649. Hiraide, M. et al.: ibid. *29*, 102 (1980)
650. Ito, S. et al.: ibid. *29*, 655 (1980)
651. Kar, K.R., Singh, G.: Mikrochim. Acta *1968*, 560 (1968)
652. Kar, K.R. et al.: ibid. *1968*, 1198 (1968)
653. Feldman, C., Rains, T.C.: Anal. Chem. *36*, 405 (1964)
654. Disam, A. et al.: Fresenius Z. Anal. Chem. *295*, 97 (1979)
655. Gregorowicz, Z. et al.: ibid. *303*, 381 (1980)
656. Hoshino, Y. et al.: J. Chem. Soc. Jpn. *1977*, 808 (1977); *1981*, 19 (1981)
657. Wolff, E.W. et al.: Anal. Chem. *53*, 1566 (1981)

658. Sebba, F.: Ion Flotation, Amsterdam, Elsevier 1962
659. Lemlich, R. (ed.): Adsorptive Bubble Separation Techniques, New York, Academic Press 1972
660. Grieves, R.B.: Chem. Eng. J. *9*, 93 (1975)
661. Fukuda, K., Mizuike, A.: Bunseki Kagaku *17*, 319 (1968)
662. Mizuike, A. et al.: ibid. *18*, 519 (1969)
663. Mizuike, A., Hiraide, M.: Pure & Appl. Chem., *54*, 1556 (1982)
664. Hiraide, M., Mizuike, A.: Rev. Anal. Chem., in press
665. Hiraide, M., Mizuike, A.: Bunseki Kagaku *26*, 655 (1977)
666. Hiraide, M., Mizuike, A.: ibid. *29*, 84 (1980)
667. Hiraide, M., Mizuike, A.: ibid. *26*, 47 (1977)
668. Hiraide, M. et al.: Anal. Chem. *52*, 804 (1980)
669. Hiraide, M., Mizuike, A.: Bunseki Kagaku *23*, 522 (1974)
670. Mizuike, A., Hiraide, M.: Anal. Chim. Acta *69*, 231 (1974)
671. Nakashima, S.: Bull. Chem. Soc. Jpn. *52*, 1844 (1979)
672. Nakashima, S.: Anal. Chem. *51*, 654 (1979)
673. Nakashima, S.: Fresenius Z. Anal. Chem. *303*, 10 (1980)
674. Nakashima, S.: Bull. Chem. Soc. Jpn. *54*, 291 (1981)
675. Kim, Y.S., Zeitlin, H.: Separ. Sci. *6*, 505 (1971)
676. Chaine, F.E., Zeitlin, H.: ibid. *9*, 1 (1974)
677. Hagadone, M., Zeitlin, H.: Anal. Chim. Acta *86*, 289 (1976)
678. Tzeng, J.H., Zeitlin, H.: ibid. *101*, 71 (1978)
679. Kim, Y.S., Zeitlin, H.: Anal. Chem. *43*, 1390 (1971)
680. Kim, Y.S., Zeitlin, H.: Separ. Sci. 7, 1 (1972)
681. Nakashima, S.: Analyst *103*, 1031 (1978)
682. Nakashima, S.: Bunseki Kagaku *28*, 561 (1979)
683. Hiraide, M. et al.: Anal. Chim. Acta *81*, 185 (1976)
684. Leung, G. et al.: ibid. *60*, 229 (1972)
685. Williams, W.J., Gillam, A.H.: Analyst *103*, 1239 (1978)
686. Voyce, D., Zeitlin, H.: Anal. Chim. Acta *69*, 27 (1974)
687. Rothstein, N., Zeitlin, H.: Anal. Lett. *9*, 461 (1976)
688. Hiraide, M., Mizuike, A.: Bull. Chem. Soc. Jpn. *48*, 3753 (1975)
689. Hiraide, M. et al.: J. Chem. Soc. Jpn. *1981*, 161 (1981)
690. Mizuike, A. et al.: Bunseki Kagaku *26*, 137 (1977)
691. Hiraide, M., Mizuike, A.: Talanta *22*, 539 (1975)
692. Mizuike, A. et al.: Bunseki Kagaku *26*, 72 (1977)
693. Sekine, K., Onishi, H.: Anal. Chim. Acta *62*, 468 (1972)
694. Sekine, K.: Mikrochim. Acta *1975 I*, 313 (1975)
695. Aoyama, M. et al.: Bunseki Kagaku *30*, 224 (1981)
696. Aoyama, M. et al.: Anal. Chim. Acta *129*, 237 (1981)
697. Hobo, T. et al.: Bunseki Kagaku *24*, 288 (1975)
698. Hobo, T. et al.: ibid. *27*, 104 (1978)
699. Suzuki, S. et al.: ibid. *31*, 13 (1982)
700. Aoyama, M. et al.: ibid. *31*, E7 (1982)
701. Aoyama, M. et al.: ibid. *31*, E99 (1982)
702. Kotsuji, K. et al.: ibid. *26*, 475 (1977)
703. Kotsuji, K. et al.: ibid. *28*, 263 (1979)
704. Shapiro, J.: Science *133*, 2063 (1961)
705. Kobayashi, S., Lee, G.F.: Anal. Chem. *36*, 2197 (1964)
706. Smith, G.H., Tasker, M.P.: Anal. Chim. Acta *33*, 559 (1965)
707. Shapiro, J.: Anal. Chem. *39*, 280 (1967)
708. Mizuike, A. Kano, S.: Bunseki Kagaku *17*, 354 (1968)
709. Murozumi, M. et al.: ibid. *19*, 1057 (1970)
710. Yonehara, N., Kamada, M.: ibid. *26*, 129 (1977); *30*, 620 (1981)
711. Yonehara, N. et al.: ibid. *30*, 617 (1981)

712. Yonehara, N. et al.: J. Chem. Soc. Jpn. *1981*, 166 (1981)
713. Pfann, W.G.: Zone Melting, New York, Wiley 1966[2]
714. Pfann, W.G., Theuerer, H.C.: Anal. Chem. *32*, 1574 (1960)
715. Stumm, W., Brauner, P.A.: Chemical Speciation, in: Chemical Oceanography (ed.) Riley, J.P., Skirrow, G., p. 173, New York, Academic Press 1975[2]
716. Florence, T.M.: Water Res. *11*, 681 (1977)
717. Florence, T.M., Batley, G.E.: Talanta *24*, 151 (1977)
718. Stumm, W., Morgan, J.J.: Aquatic Chemistry, New York, Wiley 1981[2]
719. Buffle, J.: Tr. Anal. Chem. *1*, 90 (1981)
720. Smits, J. et al.: Anal. Chim. Acta *111*, 215 (1979)
721. Bruland, K.W. et al.: ibid. *105*, 233 (1979)
722. Lieser, K.H. et al.: Mikrochim. Acta *1980 II*, 445 (1980)
723. Batley, G.E., Gardner, D.: Water Res. *11*, 745 (1977)
724. Beneš, P. et al.: ibid. *10*, 711 (1976)
725. Slowey, J.F., Hood, D.W.: Geochim. Cosmochim. Acta *35*, 121 (1971)
726. Beneš, P., Steinnes, E.: Water Res. *8*, 947 (1974)
727. Bender, M.E. et al.: Environ. Sci. Technol. *4*, 520 (1970)
728. Mantoura, R.F.C., Riley, J.P.: Anal. Chim. Acta *78*, 193 (1975)
729. Means, J.L. et al.: Limnol. Oceanogr. *22*, 957 (1977)
730. Steinberg, C.: Water Res. *14*, 1239 (1980)
731. Abdullah, M.I. et al.: Anal. Chim. Acta *84*, 363 (1976)
732. Andreae, M.O.. Anal. Chem. *49*, 820 (1977)
733. Aggett, J., Aspell, A.C.: Analyst *101*, 341 (1976)
734. Howard, A.G., Arbab-Zavar, M.H.: ibid. *105*, 338 (1980)
735. Hinners, T.A.: ibid. *105*, 751 (1980)
736. Braman, R.S. et al.: Anal. Chem. *49*, 621 (1977)
737. Shaikh, A.U., Tallman, D.E.: Anal. Chim. Acta *98*, 251 (1978)
738. Howard, A.G., Arbab-Zavar, M.H.: Analyst *106*, 213 (1981)
739. Cutter, G.A.: Anal. Chim. Acta *98*, 59 (1978)
740. Hodge, V.F. et al.: Anal. Chem. *51*, 1256 (1979)
741. Andreae, M.O. et al.: ibid. *53*, 1766 (1981)
742. Umezaki, Y., Iwamoto, K.: Bunseki Kagaku *20*, 173 (1971)
743. Baltisberger, R.J., Knudson, C.L.: Anal. Chim. Acta *73*, 265 (1974)
744. Minagawa, K. et al.: ibid. *115*, 103 (1980)
745. Oda, C.E., Ingle, Jr., J.D.: Anal. Chem. *53*, 2305 (1981)
746. Mizunuma, H. et al.: Bunseki Kagaku *28*, 695 (1979)
747. Goulden, P.D., Anthony, D.H.J.: Anal. Chim. Acta *120*, 129 (1980)
748. Slowey, J.F. et al.: Nature *214*, 377 (1967)
749. Florence, T.M., Batley, G.E.: Talanta *23*, 179 (1976)
750. Hiiro, K. et al.: Bunseki Kagaku *25*, 122 (1976)
751. Osaki, S. et al.: ibid. *25*, 358 (1976)
752. Bergmann, H., Hardt, K.: Fresenius Z. Anal. Chem. *297*, 381 (1979)
753. Jong, G.J. de, Brinkman, U.A.Th.: Anal. Chim. Acta *98*, 243 (1978)
754. Kamada, T.: Talanta *23*, 835 (1976)
755. Yamamoto, Y., Kamada, T.: Bunseki Kagaku *25*, 567 (1976)
756. Chakraborti, D. et al.: Anal. Chim. Acta *120*, 121 (1980)
757. Kamada, T. et al.: Talanta *25*, 15 (1978)
758. Shimoishi, Y., Tôei, K.: Anal. Chim. Acta *100*, 65 (1978)
759. Measures, C.I., Burton, J.D.: ibid. *120*, 177 (1980)
760. Chuecas, L., Riley, J.P.: ibid. *35*, 240 (1966)
761. Pik, A.J. et al.: ibid. *124*, 351 (1981)
762. Yamazaki, H.. ibid. *113*, 131 (1980)
763. Yoshii, O. et al.: Bunseki Kagaku *26*, 91 (1977)
764. Muzzarelli, R.A.A., Rocchetti, R.: Anal. Chim. Acta *69*, 35 (1974)
765. Bruland, K.W. et al.: ibid. *105*, 233 (1979)

766. Figura, P., McDuffie, B.: Anal. Chem. *52*, 1433 (1980)
767. Sugimura, Y. et al.: Deep-Sea Res. *25*, 309 (1978)
768. Sugimura, Y. et al.: J. Oceanogr. Soc. Jpn. *34*, 93 (1978)
769. Grabinski, A.A.: Anal. Chem. *53*, 966 (1981)
770. Pacey, G.E., Ford, J.A.: Talanta *28*, 935 (1981)
771. Koshima, H., Onishi, H.: Bunseki Kagaku *30*, 672 (1981)
772. Pankow, J.F., Janauer, G.E.: Anal. Chim. Acta *69*, 97 (1974)
773. Schwedt, G.: Fresenius Z. Anal. Chem. *295*, 382 (1979)
774. Fukai, R., Vas, D.: J. Oceanogr. Soc. Jpn. *23*, 298 (1967)
775. Cranston, R.E., Murray, J.W.: Anal. Chim. Acta *99*, 275 (1978)
776. Nakayama, E. et al.: ibid. *130*, 289, 401 (1981); *131*, 247 (1981)
777. Miyazaki, A., Barnes, R.M.: Anal. Chem. *53*, 364 (1981)
778. Robberecht, H.J., Van Grieken, R.E.: ibid. *52*, 449 (1980)
779. Sugimura, Y., Suzuki, Y.: J. Oceanogr. Soc. Jpn. *33*, 23 (1977)
780. Cadle, R.D.: The Measurement of Airborne Particles, New York, Wiley 1975
781. Malissa, H. (ed.): Analysis of Airborne Particles by Physical Methods, West Palm Beach, CRC Press 1978
782. Paez, D.M., Guagnini, O.A.: Mikrochim. Acta *1971*, 220 (1971)
783. Vasireddy, S. et al.: Anal. Chem. *53*, 868 (1981)
784. Kashihira, N. et al.: Bunseki Kagaku *31*, E 13 (1982)
785. Janssen, J.H. et al.: Anal. Chim. Acta *92*, 71 (1977)

Appendix

A. 1 Solvents

Table 43. Physical properties of solvents

Solvent	Relative dielectric constant	Boiling point [°C]	Density [g cm^{-3}]	Solubility [wt%]	
				Solvent in water	Water in solvent
Hexane	1.9	69	0.66	0.001	0.01
Cyclohexane	2.0	81	0.77	0.01	0.006
Carbon tetrachloride	2.2	77	1.58	0.077	0.01
p-Dioxane	2.2	101	1.03	Water-miscible	
Benzene	2.3	80	0.87	0.178	0.06
Toluene	2.4	111	0.86	0.05	0.03
m-Xylene	2.4	139	0.86	0.02	0.04
Carbon disulfide	2.6	46	1.26	0.29	< 0.005
Pentyl ether	2.8	187	0.78	–	–
Isopentyl ether	2.8	173	0.78	0.02	–
Butyl ether	3.1	142	0.76	0.03	0.19
Propyl ether	3.4	90	0.74	0.49	0.45
Isopropyl ether	3.9	68	0.72	1.2	0.57
Ethyl ether	4.3	35	0.71	6.04	1.47
Pentyl acetate	4.8	149	0.87	0.17	1.15
Chloroform	4.8	61	1.48	0.82	0.07
Isobutyl acetate	5.3	118	0.87	0.67	1.64
Ethyl acetate	6.0	77	0.90	8.08	2.94
Methyl acetate	6.7	56	0.93	24	8
Tetrahydrofuran	7.6	66	0.89	Water-miscible	
Tributyl phosphate (TBP)	8.0	289	0.98	0.04	4.67
Isobutylmethyl ketone (MIBK)	13.1	117	0.80	1.7	1.9
3-Methyl-1-butanol	14.7	131	0.81	2.67	9.61
Cyclohexanol	15.0	161	0.97	3.75	11.78
Methyl cellosolve (2-Methoxyethanol)	16.9	125	0.96	Water-miscible	
1-Butanol	17.5	118	0.81	7.45	20.5
2-Methyl-1-propanol	17.9	108	0.80	10	16.9
Cyclohexanone	18.3	156	0.94	2.3	8.0
Acetone	20.7	56	0.78	Water-miscible	
Bis(2-chloroethyl) ether	21.2	179	1.21	1.02	0.1
Ethanol	24.6	78	0.79	Water-miscible	
Methanol	32.7	65	0.79	Water-miscible	
Nitrobenzene	34.8	211	1.20	0.19	0.24
Ethylene glycol (1,2-Ethanediol)	37.7	197	1.11	Water-miscible	
Water	78.4	100	1.00	–	–

[Data from Riddick, J.A., Bunger, W.B.: Organic Solvents, New York, Wiley-Interscience 1970[3]]

A. 2 Masking Agents

Table 44. Masking agents for cations

Cation	Masking agents[a]
Ag	CN^-, I^-, Br^-, Cl^-, SCN^-, $S_2O_3^{2-}$, NH_3, thiourea, TGA, DDTC
Al	OH^-, F^-, BF_4^-, acetate, citrate, formate, oxalate, salicylate, tartrate, acetylacetone, mannitol, EDTA, TEA, tiron, SSA, DMP
As	S^{2-}, DMP, unithiol, citrate, tartrate, OH^-
Au	CN^-, Br^-, $S_2O_3^{2-}$
Ba	SO_4^{2-}, F^-, PO_4^{3-}, EDTA, DHG, citrate, tartrate
Be	Citrate, tartrate, acetylacetone, SSA, tiron, EDTA, F^-
Bi	I^-, SCN^-, $S_2O_3^{2-}$, Cl^-, F^-, OH^-, DDTC, DMP, dithizone, TGA, unithiol, cysteine, thiourea, citrate, tartrate, oxalate, tiron, SSA, NTA, EDTA, TEA, DHG, triphosphate, ascorbic acid
Ca	F^-, BF_4^-, polyphosphate, oxalate, tartrate, NTA, EDTA, DHG
Cd	I^-, CN^-, $S_2O_3^{2-}$, SCN^-, cysteine, DDTC, DMP, dithizone, TGA, unithiol, citrate, tartrate, glycine, DHG, NTA, EDTA, NH_3, tetren, phen
Ce	F^-, PO_4^{3-}, $P_2O_7^{4-}$, citrate, tartrate, DHG, NTA, EDTA, tiron
Co	CN^-, SCN^-, $S_2O_3^{2-}$, F^-, NO_2^-, citrate, tartrate, tiron, glycine, DHG, TEA, EDTA, DDTC, DMP, TGA, unithiol, NH_3, en, tren, tetren, penten, phen, DMG, H_2O_2, triphosphate
Cr	Formate, acetate, citrate, tartrate, tiron, SSA, glycerol, DHG, NTA, EDTA, TEA, F^-, PO_4^{3-}, $P_2O_7^{4-}$, triphosphate
Cu	NH_3, en, tren, trien, tetren, penten, phen, glycine, DHG, tartrate, citrate, tiron, NTA, EDTA, S^{2-}, DDTC, DMP, TGA, cysteine, thiourea, CN^-, $S_2O_3^{2-}$, $SCN^- + SO_3^{2-}$
Fe	F^-, PO_4^{3-}, $P_2O_7^{4-}$, OH^-, CN^-, tartrate, oxalate, citrate, glycerol, NTA, EDTA, TEA, acetylacetone, tiron, DHG, SSA, S^{2-}, $S_2O_3^{2-}$, DMP, TGA, thiourea, unithiol, phen, tren
Ga	OH^-, Cl^-, citrate, oxalate, tartrate, EDTA, SSA, unithiol
Ge	F^-, oxalate, tartrate
Hf	PO_4^{3-}, $P_2O_7^{4-}$, F^-, SO_4^{2-}, citrate, oxalate, tartrate, NTA, EDTA, DHG, SSA, TEA, H_2O_2
Hg	Cl^-, CN^-, I^-, SCN^-, $S_2O_3^{2-}$, SO_3^{2-}, cysteine, DDTC, DMP, TGA, thiourea, unithiol, tartrate, citrate, NTA, EDTA, TEA, DHG, trien, tren, tetren, penten, phen
In	F^-, Cl^-, SCN^-, TGA, thiourea, unithiol, tartrate, EDTA, TEA
Ir	CN^-, SCN^-, citrate, tartrate, thiourea
Mg	OH^-, F^-, BF_4^-, $P_2O_7^{4-}$, PO_4^{3-}, hexametaphosphate, citrate, tartrate, oxalate, tiron, glycol, NTA, EDTA, TEA, DHG
Mn	F^-, $P_2O_7^{4-}$, CN^-, triphosphate, citrate, tartrate, oxalate, tiron, SSA, NTA, EDTA, TEA, DHG, DMP, S^{2-}, tren, phen
Mo	Citrate, tartrate, oxalate, acetylacetone, tiron, NTA, EDTA, DHG, F^-, triphosphate, H_2O_2, SCN^-, mannitol, ascorbic acid
Nb	F^-, OH^-, citrate, tartrate, oxalate, tiron, H_2O_2
Ni	CN^-, F^-, SCN^-, NH_3, en, tren, tetren, penten, phen, citrate, tartrate, NTA, EDTA, SSA, DHG, glycine, DDTC, TGA, DMG, triphosphate

Table 44 (continued)

Cation	Masking agents[a]
Os	CN^-, SCN^-, thiourea
Pb	OH^-, F^-, Cl^-, I^-, SO_4^{2-}, $S_2O_3^{2-}$, PO_4^{3-}, triphosphate, DDTC, DMP, TGA, unithiol, acetate, citrate, tartrate, tiron, NTA, EDTA, TEA, DHG
Pd	CN^-, SCN^-, thiourea, I^-, NO_2^-, $S_2O_3^{2-}$, NH_3, citrate, tartrate, NTA, EDTA, TEA, DHG, acetylacetone
Pt	CN^-, SCN^-, I^-, NO_2^-, $S_2O_3^{2-}$, NH_3, citrate, tartrate, NTA, EDTA, thiourea
Rare earths	F^-, citrate, tartrate, oxalate, EDTA
Rh	Citrate, tartrate, thiourea
Sb	F^-, Cl^-, I^-, OH^-, S^{2-}, $S_2O_3^{2-}$, citrate, tartrate, oxalate, TEA, DMP, unithiol, TGA
Se	F^-, I^-, S^{2-}, SO_3^{2-}, tartrate, citrate
Sn	F^-, Cl^-, I^-, OH^-, PO_4^{3-}, oxalate, citrate, tartrate, EDTA, TEA, DMP, TGA, unithiol
Sr	F^-, SO_4^{2-}, citrate, tartrate, NTA, EDTA, DHG
Ta	F^-, OH^-, citrate, tartrate, oxalate, EDTA, H_2O_2
Te	F^-, I^-, S^{2-}, SO_3^{2-}, citrate, tartrate
Th	F^-, SO_4^{2-}, acetate, oxalate, citrate, tartrate, SSA, TEA, DHG, NTA, EDTA, tiron, acetylacetone
Ti	F^-, OH^-, SO_4^{2-}, PO_4^{3-}, triphosphate, citrate, tartrate, SSA, TEA, DHG, NTA, EDTA + H_2O_2, tiron, mannitol, ascorbic acid, ferron, H_2O_2
Tl	Cl^-, CN^-, citrate, tartrate, oxalate, TEA, DHG, NTA, EDTA, TGA, unithiol
U	$(NH_4)_2CO_3$, citrate, tartrate, oxalate, acetylacetone, F^-, PO_4^{3-}, EDTA, H_2O_2, phen, SSA
V	Citrate, tartrate, oxalate, TEA, tiron, mannitol, EDTA, CN^-, H_2O_2
W	F^-, PO_4^{3-}, SCN^-, citrate, tartrate, oxalate, tiron, mannitol, EDTA, H_2O_2, triphosphate
Zn	CN^-, OH^-, SCN^-, triphosphate, citrate, tartrate, glycol, glycerol, TEA, NTA, EDTA, NH_3, tren, tetren, penten, phen, glycine, DHG, DMP, dithizone, TGA, unithiol
Zr	F^-, CO_3^{2-}, PO_4^{3-}, $P_2O_7^{4-}$, SO_4^{2-} + H_2O_2, OH^-, citrate, tartrate, oxalate, salicylate, SSA, pyrogallol, tiron, TEA, DHG, NTA, EDTA, cysteine

[a] Abbreviations: DDTC, diethyldithiocarbamate; DHG, *N,N*-dihydroxyethylglycine; DMG, dimethylglyoxime; DMP, 2,3-dimercaptopropanol; EDTA, ethylenediaminetetraacetic acid; en, ethylenediamine; NTA, nitrilotriacetic acid; penten, pentaethylenehexamine; phen, 1,10-phenanthroline; SSA, sulfosalicylic acid; TEA, triethanolamine; tetren, tetraethylenepentamine; TGA, thioglycolic acid; tren, triaminotriethylamine; trien, triethylenetetramine
[From Perrin, D.D.: Masking and Demasking in Analytical chemistry in: Treatise on Analytical Chemistry (ed.) Kolthoff, I.M., Elving, P.J., Part I, Vol. 2, p. 609, New York, Wiley 1979[2]]

A. 3 Ion Exchange Data

Table 45. Distribution coefficients (D) with strong-acid cation exchange resin (AG50W-X8) in hydrochloric acid

Cation	HCl						
	0.1 N	0.2 N	0.5 N	1.0 N	2.0 N	3.0 N	4.0 N
ZrO^{2+}	$> 10^5$	$> 10^5$	$\sim 10^5$	7250	489	61	14.5
Th^{4+}	$> 10^5$	$> 10^5$	$\sim 10^5$	2049	239	114	67
La^{3+}	$> 10^5$	10^5	2480	265.1	48	18.8	10.4
Ce^{3+}	$> 10^5$	10^5	2460	264.8	48	18.8	10.5
Y^{3+}	$> 10^5$	$> 10^4$	1460	144.6	29.7	13.6	8.6
Ba^{2+}	$> 10^4$	2930	590	126.9	36	18.5	11.9
Hg^{+a}	$> 10^4$	7600	640	94.2	33	19.2	13.6
Al^{3+}	8200	1900	318	60.8	12.5	4.7	2.8
Sr^{2+}	4700	1070	217	60.2	17.8	10.0	7.5
Ga^{3+}	$> 10^4$	3036	260	42.58	7.75	3.2	0.36
Ca^{2+}	3200	790	151	42.29	12.2	7.3	5.0
Pb^{2+a}	$> 10^4$	1420	183	35.66	9.8	6.8	4.5
Fe^{3+}	9000	3400	225	35.45	5.2	3.6	2.0
Cr^{3+}	1130	262	73	26.69^b	7.9	4.8	2.7
Tl^{+a}	173	91	41	22.32	9.9	5.8	3.3
Ni^{2+}	1600	450	70	21.85	7.2	4.7	3.1
Co^{2+}	1650	460	72	21.29	6.7	4.2	3.0
Mg^{2+}	1720	530	88	20.99	6.2	3.5	3.5
Mn^{2+}	2230	610	84	20.17	6.0	3.9	2.5
Fe^{2+}	1820	370	66	19.77	4.1	2.7	1.8
Cs^+	182	99	44	19.41	10.4	. . .	. . .
UO_2^{2+}	5460	860	102	19.20	7.3	4.9	3.3
Ag^{+a}	156	83	35	18.08	7.9	5.4	4.0
Cu^{2+}	1510	420	65	17.50	4.3	2.8	1.8
Hg^{2+a}	4700	1090	121	16.85	5.9	3.9	2.8
Zn^{2+}	1850	510	64	16.03	3.7	2.4	1.6
Rb^+	120	72	33	15.43	8.1	. . .	. . .
K^+	106	64	29	13.87	7.4	. . .	. . .
Be^{2+}	255	117	42	13.33	5.2	3.3	2.4
Ti^{4+}	$> 10^4$	297	39	11.86	3.7	2.4	1.7
V^{4+}		230	44	7.20	. . .	. . .	. . .
Na^+	52	28.3	12	5.59	3.6	. . .	. . .
Li^+	33	18.9	8.1	3.83	2.5	. . .	. . .
Sn^{4+}	$\sim 10^4$	45	6.2	1.60	1.2	. . .	. . .
Cd^{2+}	510	84	6.5	1.54	1.0	0.6	. . .
V^{5+}	13.9	7.0	5.0	1.10	0.7	0.2	0.3
Mo^{5+}	10.9	4.5	0.3	0.81	0.2	0.4	0.3
Se^{4+}	1.1	0.6	0.8	0.63	1.0	. . .	0.7
Bi^{3+}	Ppt.	Ppt.	< 1.0	1.0	1.0	1.0	1.0
As^{3+}	1.4	1.6	2.2	3.81	2.2	. . .	. . .
Sb^{3+}	Ppt.	Ppt.	Ppt.	Ppt.	2.8	. . .	. . .
Pt^{4+}	. . .	. . .	. . .	1.4	. . .	. . .	. . .
Au^{3+}	0.5	0.1	0.4	0.84	1.0	0.7	0.2
Hg^{2+}	1.6	0.9	0.5	0.28	0.3	0.2	0.2

a   Done in nitric acid
b   More than one cationic species present

[Reprinted with permission from Strelow, F.W.E.: Anal. Chem. *32*, 1186 (1960) Copyright 1960 American Chemical Society]

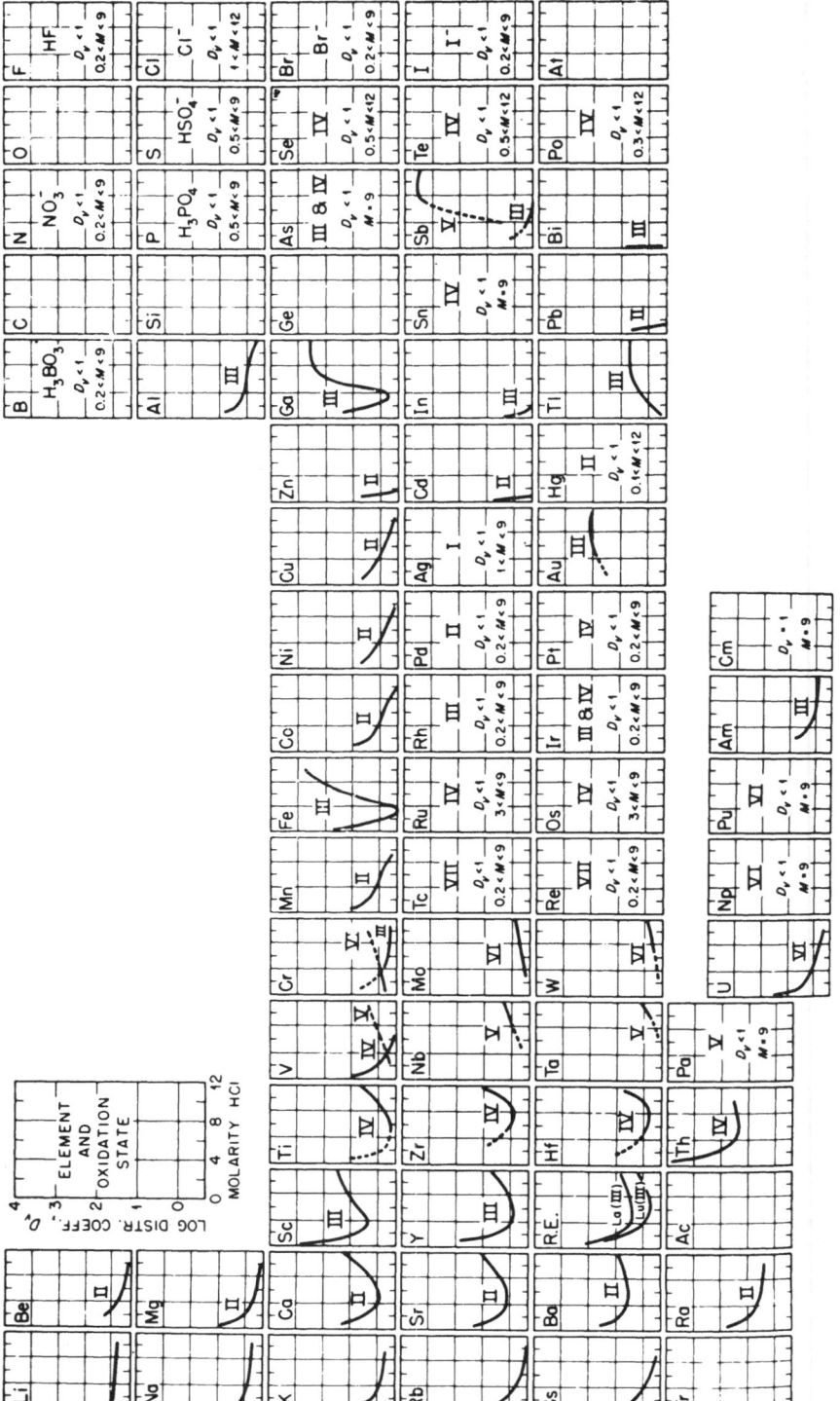

Fig. 37. Distribution coefficients (D_v) with strong-acid cation exchange resin (Dowex 50-X4) in hydrochloric acid [Reprinted with permission from Nelson, F., Murase, T., Kraus, K.A.: J. Chromatog. *13*, 532 (1964) Copyright 1964 Elsevier Scientific Publishing Company]

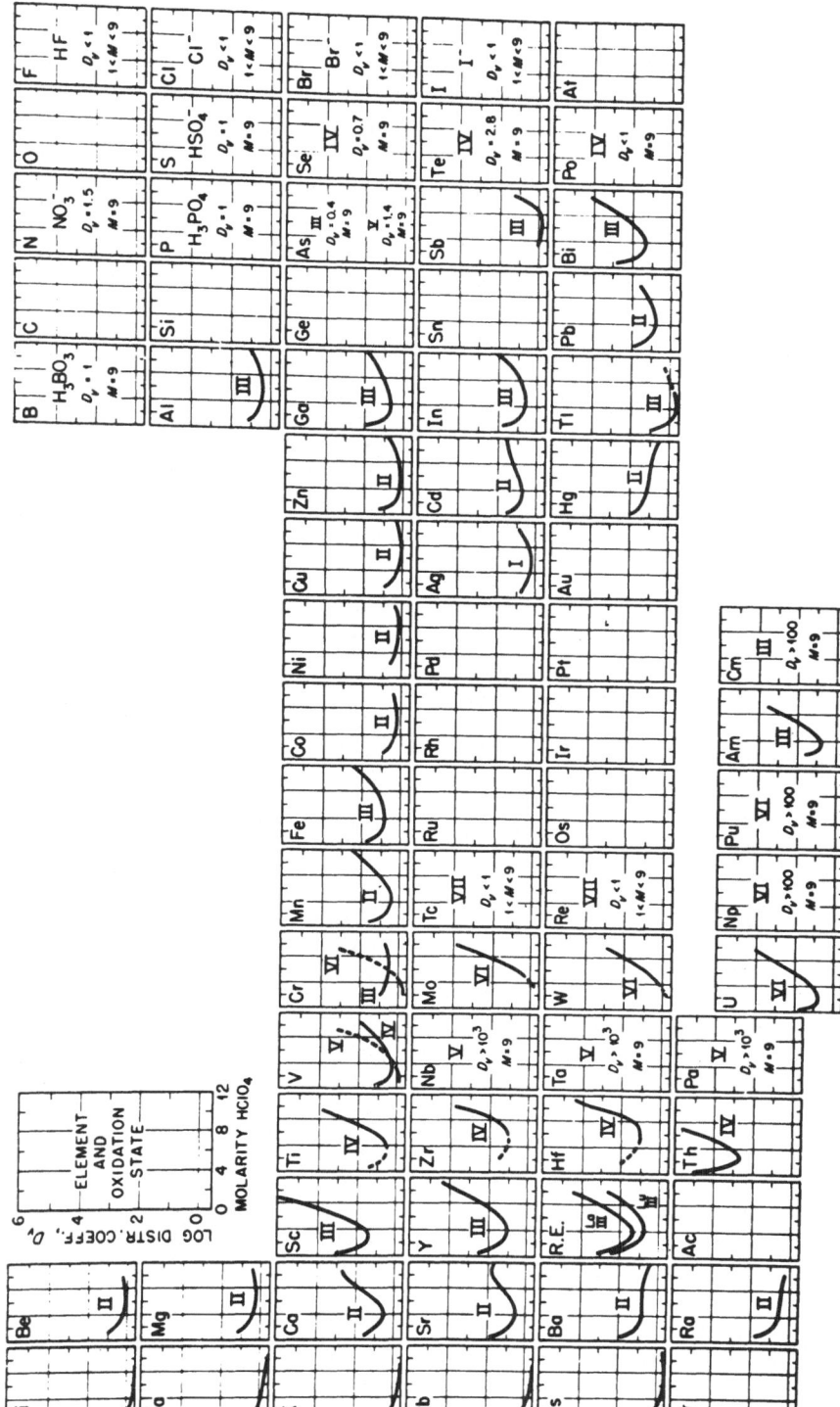

Fig. 38. Distribution coefficients (D_V) with strong-acid cation exchange resin (Dowex 50-X4) in perchloric acid [Reprinted with permission from Nelson, F., Murase, T., Kraus, K.A.: J. Chromatog. *13*, 533 (1964) Copyright 1964 Elsevier Scientific Publishing Company]

Table 46. Distribution coefficients (D) with strong-acid cation exchange resin (AG50W-X8) in perchloric acid

Element	$HClO_4$						
	0.1 M	0.2 M	0.5 M	1.0 M	2.0 M	3.0 M	4.0 M
Zr(IV)	$> 10^4$	$> 10^4$	$> 10^4$	$> 10^4$	1960	593	333
Th(IV)	$> 10^4$	$> 10^4$	$> 10^4$	5780	844	509	686
La(III)	$> 10^4$	$> 10^4$	2470	475	118	65	58
Ce(III)	$> 10^4$	$> 10^4$	2380	459	114	58	53
Dy(III)	$> 10^4$	$> 10^4$	1370	258	63	45.7	39.1
Y(III)	$> 10^4$	$> 10^4$	1390	246	59	31.1	23.7
Bi(III)	$> 10^4$	$> 10^4$	935	243	76	46.5	41.9
Yb(III)	$> 10^4$	$> 10^4$	1120	205	51	25.7	19.6
Tl(III)	2260	1550	548	176	57	37.9	40.6
Hg(I)	$> 10^4$	4160	548	147	37.5	16.2	9.0
In(III)	$> 10^4$	6620	619	128	31.6	17.1	13.9
Ba(III)	8350	2280	429	127	44.3	25.1	19.0
Cr(III)	$> 10^4$	8410	585	120	29.0	13.6	10.5
Fe(III)	$> 10^4$	7470	562	119	29.9	15.5	12.2
Pb(II)	6670	1850	368	117	38.9	21.9	17.1
Ga(III)	$> 10^4$	5870	556	112	29.0	14.6	10.8
Al(III)	$> 10^4$	5250	516	106	30.7	16.0	11.1
Hg(II)	2920	937	222	85	38.4	26.1	22.9
Sr(II)	2850	870	198	67	23.5	13.4	10.4
Ca(II)	1910	639	147	49.6	17.5	10.4	7.7
Cd(II)	1280	423	101	35.8	13.0	8.5	6.3
Fe(II)	1160	389	95	31.9	11.8	7.0	5.2
Mn(II)	1130	387	94	31.8	11.2	6.7	4.7
Ni(II)	1120	387	91	31.5	11.0	6.7	5.3
Co(II)	1120	378	92	31.1	11.0	6.7	4.8
Cu(II)	1110	378	90	30.3	10.1	6.1	4.5
Zn(II)	1080	361	88	29.5	9.9	6.6	5.3
U(VI)	763	276	75	29.0	14.5	14.0	18.1
Sn(IV)	ppt	ppt	ppt	ppt	11.9	9.0	7.5
Mg(II)	901	312	76	24.4	7.9	4.4	3.1
Tl(I)	235	131	52	23.4	8.2	4.1	2.7
Ag(I)	161	90	39.2	20.2	10.4	7.3	5.8
Ti(IV)	1911	549	88	19.1	7.6	5.6	5.7
V(IV)	542	201	50	17.6	6.9	4.7	4.4
Be(II)	567	206	49.2	14.0	4.4	2.3	1.9
Cs(I)	124	70	27.3	12.2	ppt	ppt	ppt
Rb(I)	110	61	24.4	10.9	ppt	ppt	ppt
K(I)	94	52	21.2	9.7	ppt	ppt	ppt
NH_4(I)	73	40.5	16.1	7.1	2.9	1.5	1.2
Na(I)	58	32.1	13.8	7.0	3.7	2.7	2.1
Mo(VI)	37.6	21.8	11.7	5.5	4.4	3.9	4.5
Li(I)	33.3	18.3	7.9	3.9	2.3	1.9	1.6
V(V)*	12.6	9.3	6.4	3.0	1.1	0.8	1.0
V(V)	16.3	9.8	4.5	2.2	1.3	0.7	0.8
Mo(VI)*	1.1	0.7	0.5	0.4	0.3	0.3	1.3
W(VI)*	0.3	0.4	0.3	0.4	0.2	0.2	0.4

* containing hydrogen peroxide

[Reprinted with permission from Strelow, F.W.E., Sondorp, H.: Talanta *19*, 1115 (1972) Copyright 1972 Pergamon Press Ltd.]

Table 47. Distribution coefficients (D) with strong-acid cation exchange resin (AG50W-X8) in nitric acid

Cation	HNO₃						
	$0.1\,N$	$0.2\,N$	$0.5\,N$	$1.0\,N$	$2.0\,N$	$3.0\,N$	$4.0\,N$
Zr(IV)	$> 10^4$	$> 10^4$	$> 10^4$	6500	652	112	30.7
Hf(IV)	$> 10^4$	$> 10^4$	$> 10^4$	2400	166	61	20.8
Th(IV)	$> 10^4$	$> 10^4$	$> 10^4$	1180	123	43.0	24.8
La(III)	$> 10^4$	$> 10^4$	1870	267	47.3	17.1	9.1
Ce(III)	$> 10^4$	$> 10^4$	1840	246	44.2	15.4	8.2
Yb(III)	$> 10^4$	$> 10^4$	1150	193	41.3	16.0	9.0
Er(III)	$> 10^4$	$> 10^4$	1100	182	38.2	14.9	8.0
Y(III)	$> 10^4$	$> 10^4$	1020	174	35.8	13.9	10.0
Sm(III)	$> 10^4$	$> 10^4$	1000	168	29.8	10.9	7.2
Gd(III)	$> 10^4$	$> 10^4$	1000	167	29.2	10.8	6.9
In(III)	$> 10^4$	$> 10^4$	680	118	23.0	10.1	5.8
Sc(III)	$> 10^4$	3300	500	116	23.3	11.6	7.6
Cr(III)	5100	1620	418	112	27.8	19.2	10.9
Hg(I)	$> 10^4$	7600	640	94	33.5	19.2	13.6
Ga(III)	$> 10^4$	4200	445	94	20.0	9.0	5.8
Al(III)	$> 10^4$	3900	392	79	16.5	8.0	5.4
Fe(III)	$> 10^4$	4100	362	74	14.3	6.2	3.1
Ba(II)	5000	1560	271	68	13.0	6.0	3.6
Bi(III)	$> 10^4$	7340	371	61	8.0	3.7	3.0
Sr(II)	3100	775	146	39.2	8.8	6.1	4.7
Pb(II)	$> 10^4$	1420	183	35.7	8.5	5.5	4.5
Ca(II)	1450	480	113	35.3	9.7	4.3	1.8
Cd(II)	1500	392	91	32.8	10.8	6.8	3.4
Co(II)	1260	392	91	28.8	10.1	6.1	4.7
Mn(II)	1240	389	89	28.4	11.4	7.1	3.0
Ni(II)	1140	384	91	28.1	10.3	8.6	7.3
Cu(II)	1080	356	84	26.8	8.6	4.8	3.1
Zn(II)	1020	352	83	25.2	7.5	4.6	3.6
U(VI)	659	262	69	24.4	10.7	7.4	6.6
Mg(II)	794	295	71	22.9	9.1	5.8	4.1
Tl(I)	173	91	41.0	22.3	9.9	5.8	3.3
Ag(I)	156	86	36.0	18.1	7.9	5.4	4.0
Hg(II)	4700	1090	121	16.9	5.9	3.9	2.8
Cs(I)	148	81	34.8	16.8	7.6	4.7	3.4
Be(II)	553	183	52	14.8	6.6	4.5	3.1
Ti(IV)	1410	461	71	14.6	6.5	4.5	3.4
V(IV)	495	157	35.6	14.0	4.7	3.0	2.5
Rb(I)	118	68	29.1	13.4	6.6	4.1	2.9
K(I)	99	59	26.2	11.4	5.7	3.5	2.6
Te(IV)	40.3	19.7	8.5	5.0	2.4	0.6	0.2
Pd(II)	97	62	23.5	9.1	3.4	2.7	2.5
Rh(III)	78	44.7	19.5	7.8	4.1	2.1	1.0
Na(I)	54	29.4	12.7	6.3	3.4	2.0	1.3
Li(I)	33.1	18.6	8.0	3.9	2.6	1.7	1.1
V(V)	20.0	10.9	4.9	2.0	1.2	0.8	0.5
Mo(VI)	Ppt.	5.2	2.9	1.6	1.0	0.8	0.6
Nb(V)	11.6	6.3	0.9	0.2	0.1	0.1	0.1
Se(IV)	< 0.5	< 0.5	< 0.5	< 0.5	< 0.5	< 0.5	< 0.5
As(III)	< 0.1	< 0.1	< 0.1	< 0.1	< 0.1	< 0.1	< 0.1

[Reprinted with permission from Strelow, F.W.E., Rethemeyer, R., Bothma, C.J.C.: Anal. Chem. *37*, 107 (1965) Copyright 1965 American Chemical Society]

Table 48. Distribution coefficients (D) with strong-acid cation exchange resin (AG50W-X8) in sulfuric acid

Cation	H_2SO_4 0.1 N	0.2 N	0.5 N	1.0 N	2.0 N	3.0 N	4.0 N
La(III)	> 10^4	> 10^4	1860	329	68	24.3	12.1
Ce(III)	> 10^4	> 10^4	1800	318	66	23.8	11.8
Sm(III)	> 10^4	> 10^4	1460	269	56	20.1	10.0
Y(III)	> 10^4	> 10^4	1380	253	49.9	18.0	9.4
Yb(III)	> 10^4	> 10^4	1330	249	48.1	17.3	8.8
Gd(III)	> 10^4	> 10^4	1390	246	46.6	17.9	8.9
Er(III)	> 10^4	> 10^4	1300	242	48.6	16.7	8.5
Bi(III)	> 10^4	> 10^4	6800	235	32.3	11.3	6.4
Ga(III)	> 10^4	3500	618	137	26.7	10.0	4.9
Al(III)	> 10^4	8300	540	126	27.9	10.6	4.7
Hg(II)	7900	1790	321	103	34.7	16.8	12.2
In(III)	> 10^4	3190	376	87	17.2	6.5	3.8
Mn(II)	1590	610	165	59	17.4	8.9	5.5
Fe(III)	> 10^4	2050	255	58	13.5	4.6	1.8
Cr(III)	198	176	126	55	18.7	0.9	0.2
Th(IV)	> 10^4	3900	263	52	9.0	3.0	1.8
Tl(I)	452	236	97	49.7	20.6	11.6	8.7
Tl(III)	6500	1490	205	47.4	12.0	7.2	5.2
V(IV)	1230	490	140	46.6	11.5	2.4	0.4
Ni(II)	1390	590	140	46.0	16.5	6.1	2.8
Fe(II)	1600	560	139	46.0	15.3	9.8	6.6
Cd(II)	1420	540	144	45.6	14.8	6.6	4.3
Zn(II)	1570	550	135	43.2	12.2	4.9	4.0
Co(II)	1170	433	126	42.9	14.2	6.2	5.4
Cu(II)	1310	505	128	41.5	13.2	5.7	3.7
Mg(II)	1300	484	124	41.5	13.0	5.6	3.4
Sc(III)	5600	1050	141	34.9	8.5	4.4	3.4
Be(II)	840	305	79	27.0	8.2	3.9	2.6
Cs(I)	175	108	52	24.7	9.1	4.8	3.5
Rb(I)	148	91	43.8	21.3	8.3	4.4	3.1
K(I)	138	86	41.1	19.4	7.4	3.7	2.9
Rh(III)	80	49.3	28.5	16.2	4.5	2.2	1.3
Pd(II)	109	71	32.5	13.9	6.0	3.8	2.7
Hf(IV)	2690	1240	160	12.1	1.7	1.0	0.7
U(VI)	596	118	29.2	9.6	3.2	2.3	1.8
Ti(IV)	395	225	45.8	9.0	2.5	1.0	0.4
Na(I)	81	47.7	20.1	8.9	3.7	2.6	1.7
Li(I)	48.0	28.2	11.7	5.8	3.0	1.6	1.1
Te(IV)	Ppt.	30.8	9.8	5.2	2.6	0.6	0.3
Zr(IV)	546	474	98	4.6	1.4	1.2	1.0
V(V)	27.1	15.2	6.7	2.8	1.2	0.7	0.4
Nb(V)	14.2	7.4	4.0	1.9	0.7	0.5	0.3
Mo(VI)	Ppt.	5.3	2.8	1.2	0.5	0.3	0.2
Se(IV)	< 0.5	< 0.5	< 0.5	< 0.5	< 0.5	< 0.5	< 0.5
As(III)	< 0.1	< 0.1	< 0.1	< 0.1	< 0.1	< 0.1	< 0.1

[Reprinted with permission from Strelow, F.W.E., Rethemeyer, R., Bothma, C.J.C.: Anal. Chem. *37*, 108 (1965) Copyright 1965 American Chemical Society]

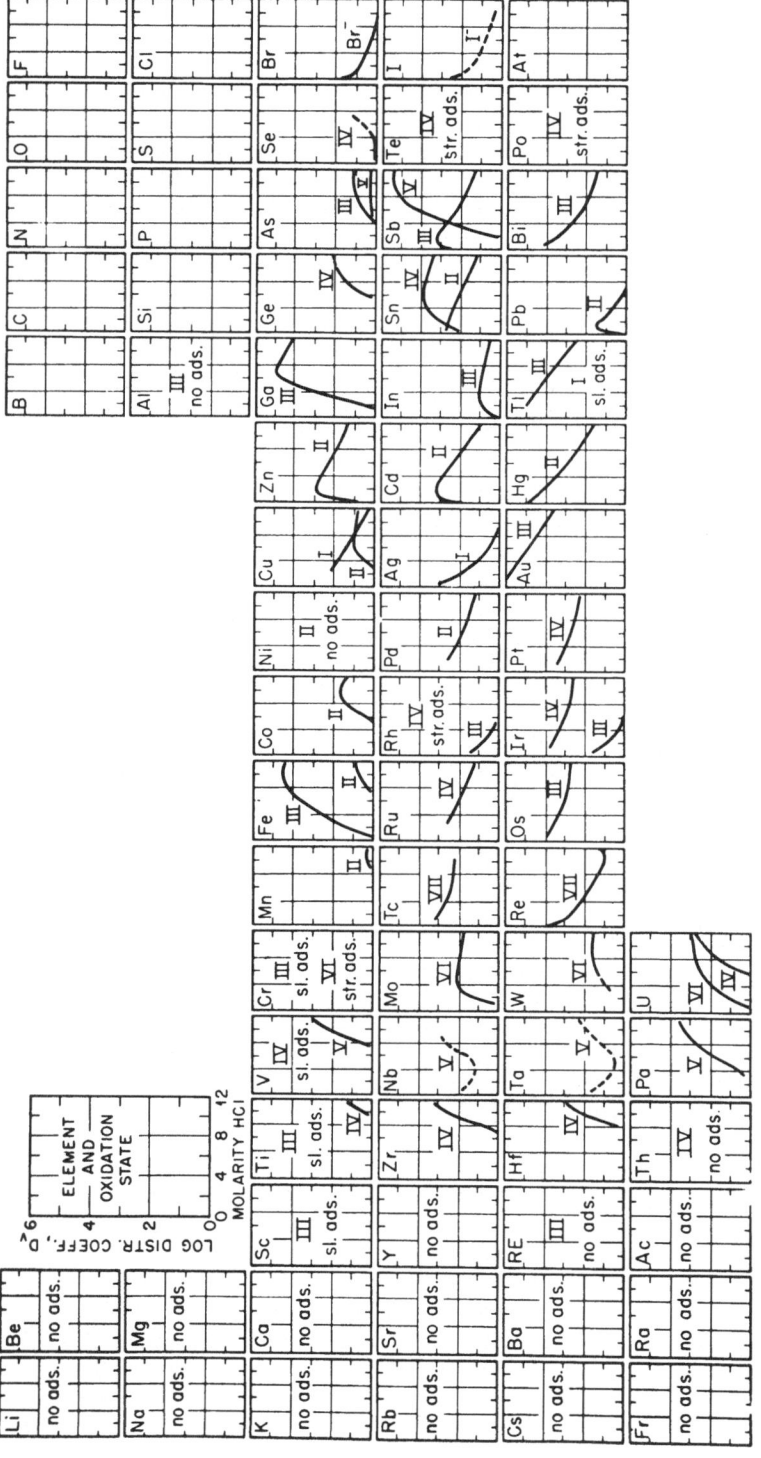

Fig. 39. Distribution coefficients (D_V) with strong-base anion exchange resin in hydrochloric acid. no ads. – no adsorption $0.1 < M \, HCl < 12$; sl. ads. – slight adsorption in $12 \, M \, HCl$ ($0.3 \leq D_V \leq 1$); str. ads. – strong adsorption $D_V \geq 1$. [Reprinted with permission from Kraus, K.A., Nelson, F.: Proc. Intern. Conf. Peaceful Uses Atomic Energy, Geneva 1955, Vol. 7, p. 118, New York, United Nations 1956]

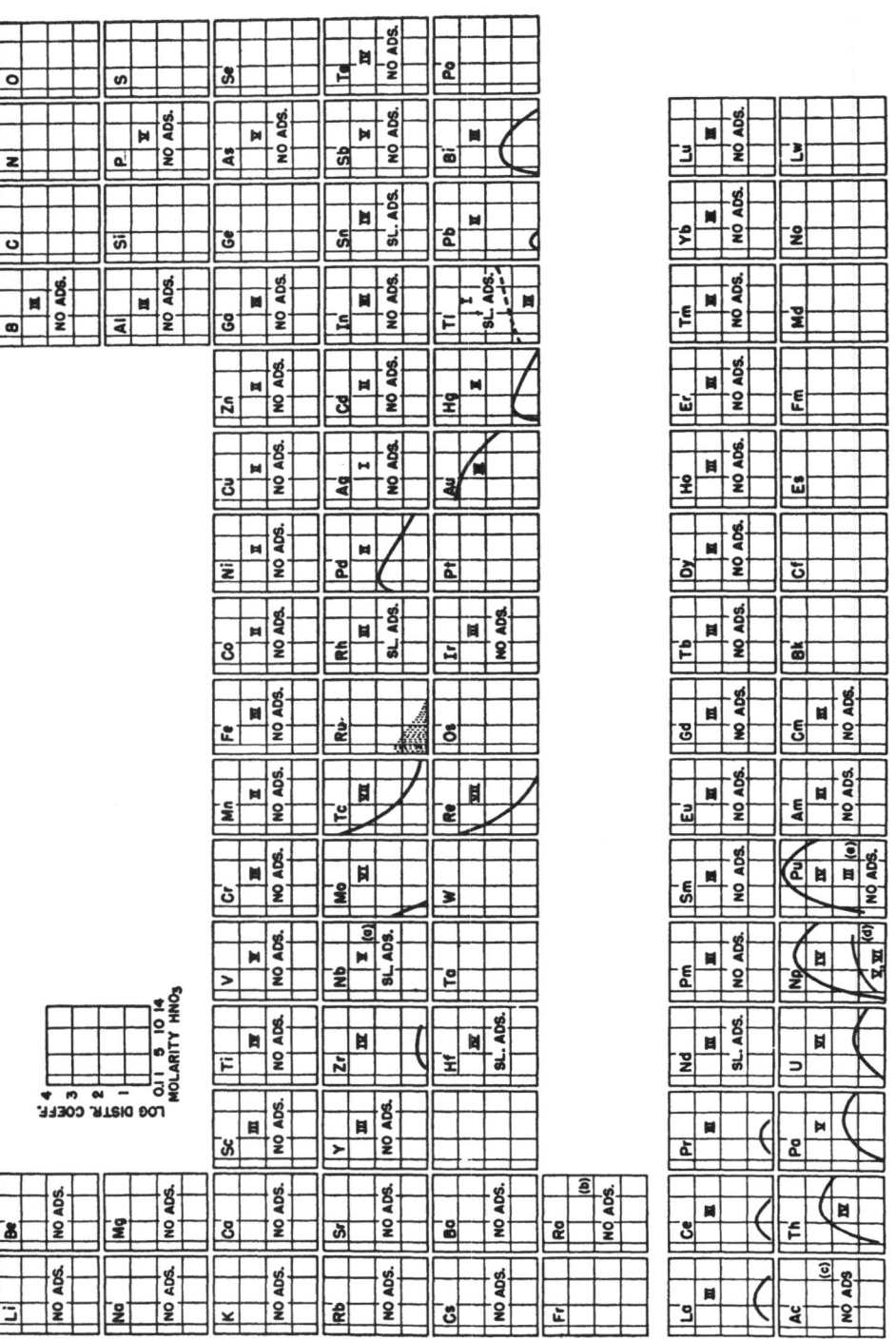

Fig. 40. Distribution coefficients (D) with strong-base anion exchange resin (Dowex 1-X10) in nitric acid. NO ADS. – no adsorption from 0.1 – 14 M HNO₃; SL. ADS. – slight adsorption. [Reprinted with permission from Faris, J.P., Buchanan, R.F.: Anal. Chem. 36, 1158 (1964) Copyright 1964 American Chemical Society]

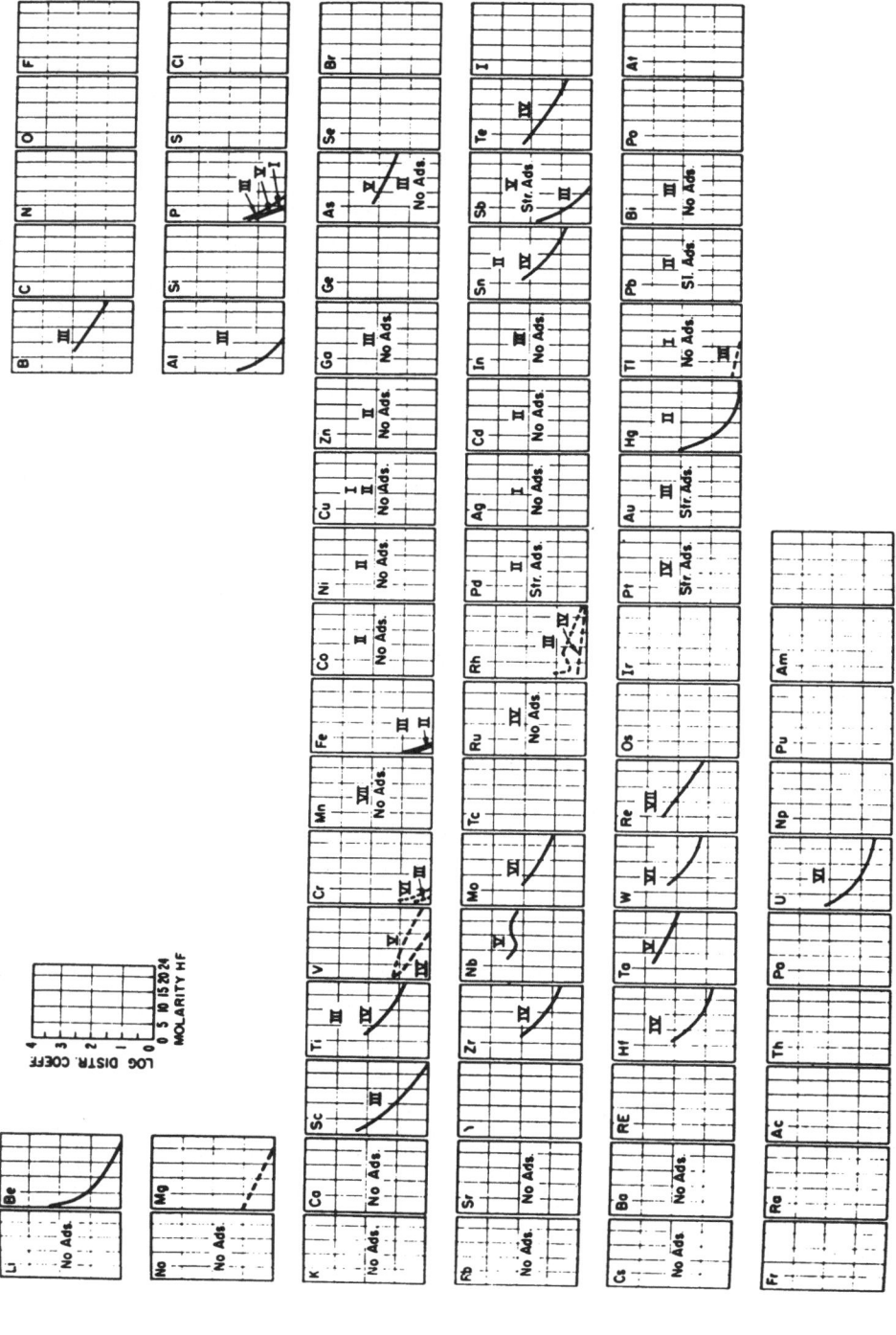

Fig. 41. Distribution coefficients (D) with strong-base anion exchange resin (Dowex 1-X10) in hydrofluoric acid. No Ads. – no adsorption from 1–24 M HF; Sl. Ads. – slight adsorption; Str. Ads. – strong adsorption: log D > 2. [Reprinted with permission from Faris, J.P.: Anal. Chem. 32, 521 (1960) Copyright 1960 American Chemical Society]

Table 49. Distribution coefficients (D) with strong-base anion exchange resin (AG1-X8) in sulfuric acid

Element	H_2SO_4								
	0.01 N	0.03 N	0.10 N	0.20 N	0.50 N	1.00 N	2.00 N	3.00 N	4.00 N
Cr(VI)	25000	18000	12000	7800	4400	2100	800	435	302
Mo(VI)	60000	527	533	671	484	232	52	13.7	4.6
W(VI)[a]	...	...	528	457	337	222	127	96	110
Mo(VI)[a]	...	...	2560	1400	451	197	74	43.3	33.0
Ir(IV)	1010	690	450	310	220	180	160	160	170
Ir(III)	625	525	388	270	218	160	118	92	75
Ta(V)[a]	...	...	1860	1070	310	138	50	11.4	3.9
Zr(IV)	hydrol	$> 10^3$	1350	704	211	47.3	11.0	5.2	2.9
U(VI)	1160	1130	521	248	91	26.6	9.3	4.8	2.9
Hf(IV)	hydrol	$> 10^3$	4700	701	57	12.0	3.2	1.9	1.2
Th(IV)	116	82	34.6	21.4	8.3	3.7	2.0	1.1	0.6
Sc(III)	64	44.5	21.5	10.9	4.8	2.6	1.5	0.9	0.5
V(V)[a]	370	102	45.4	10.9	4.6	2.5	2.1	1.9	1.9
Fe(III)	54	39.9	15.6	9.1	3.6	1.4	0.9	0.6	<0.5
Nb(V)[a]	...	...	120	96	3.4	<0.5	<0.5	<0.5	<0.5
Bi(III)			17.7	4.7	2.1	0.9	<0.5	<0.5	<0.5
Rh(III)	39.0	30.0	12.8	5.4	0.9	<0.5	<0.5	<0.5	<0.5
V(V)	1410	320	6.5	3.3	1.6	0.7	<0.5	<0.5	<0.5
In(III)	7.4	5.1	2.4	0.8	<0.5	<0.5	<0.5	<0.5	<0.5
Cr(III)	5.1	3.4	2.1	0.7	<0.5	<0.5	<0.5	<0.5	<0.5
Se(IV)	8.1	5.3	1.1	<0.5	<0.5	<0.5	<0.5	<0.5	<0.5
V(IV)	3.4	1.7	0.9	<0.5	<0.5	<0.5	<0.5	<0.5	<0.5
As(III)	2.4	1.5	0.9	0.6	<0.5	<0.5	<0.5	<0.5	<0.5
As(V)	1.3	0.8	0.6	<0.5	<0.5	<0.5	<0.5	<0.5	<0.5
Ga(III)	1.2	0.8	0.6	<0.5	<0.5	<0.5	<0.5	<0.5	<0.5
Yb(III)	1.2	0.8	0.6	0.5	0.5	0.5	0.5	0.5	<0.5
Ti(IV)[a]	hydrol	hydrol	hydrol	0.5	0.5	0.5	0.5	0.5	<0.5

a H_2O_2 present

[Reprinted with permission from Strelow, F.W.E., Bothma, C.J.C.: Anal. Chem. 39, 596 (1967) Copyright 1967 American Chemical Society]

Index of Abbreviations and Symbols

Abbreviations

AAS	atomic absorption spectrometry
AC	alternating current
APDC	ammonium pyrrolidinedithiocarbamate
	(ammonium tetramethylenedithiocarbamate)
ASV	anodic stripping voltammetry
Bu	n-butylamine
CSV	cathodic stripping voltammetry
DC	direct current
DDTC	sodium diethyldithiocarbamate
EDTA	ethylenediaminetetraacetate
EPMA	electron probe microanalysis
Fluor.	fluorometry
HEPA	high-efficiency particulate air
HPLC	high-performance liquid chromatography
ICP	inductively coupled plasma
IMA	ion probe microanalysis
MIBK	isobutylmethyl ketone
NAA	neutron activation analysis
ND	not detected
OES	optical emission spectrometry
Ox	oxine
PAN	1-(2-pyridylazo)-2-naphthol
Phot.	spectrophotometry
Polar.	polarography
SCE	saturated calomel electrode
SIMS	secondary ion mass spectrometry
SSMS	spark source mass spectrometry
TBP	tributyl phosphate
Titr.	titrimetry
Tr	traces
TTA	thenoyltrifluoroacetone
Turbid.	turbidimetry
Voltam.	voltammetry
XRF	X-ray fluorescence spectrometry

Symbols

A	analytical value. constant
B	blank value. half the band width. constant
C	impurity concentration
C_0	$C_{m,0}$ at v = 0. initial impurity concentration
C_I	concentration of element in influent
C_L	impurity concentration in liquid phase
C_{max}	maximum concentration
$C_{m,p}$	concentration of element in moving phase in plate p
C_r	total concentration of chelating agent in aqueous phase
C_S	impurity concentration in solid phase
$C_{s,p}$	concentration of element in stationary phase in plate p
D	distribution ratio. homogeneous distribution coefficient. weight distribution coefficient. diffusion constant
D_{lim}	limiting value of distribution ratio
D_v	volume distribution coefficient
E	percent extraction
F	enrichment factor
f	growth rate of solid
I	interstitial volume in column
K	equilibrium distribution coefficient
k	constant. effective distribution coefficient
K, K', K_n	stepwise formation constants
K_{ex}	extraction constant
K_H^M	selectivity coefficient of element M with respect to hydrogen ion
L	masking agent. sample length
l	molten-zone length
M	matrix. metal
N	total number of plates in column
n	number
p	ordinal number of plate
P_c	partition constant of chelate
$pH_{1/2}$	pH value where the distribution ratio is unity
P_r	partition constant of reagent
Q	quantity of element
Q_0	initial quantity of element
Q_M^0, Q_M	quantities of matrix before and after enrichment
Q_T^0, Q_T	quantities of trace element before and after enrichment
R	reagent
R_M	yield of matrix
R_T	trace recovery
r_+, r_-	radii of cation and anion
S	total volume of stationary phase in column
T	trace element
$T_{n,r}$	fraction of solute in $L_r - U_{n-r}$ pair

t	time
V	volume of aqueous phase. volume of eluent or effluent
v	$V/(V_m + D_v V_s)$
V_m	volume of moving phase in plate
V_o	volume of organic phase
V_R	retention volume
V_s	volume of stationary phase in plate
X	halide
x	distance
z_+, z_-	electric charges of cation and anion
α	side reaction coefficient
β	overall formation constant
δ	thickness of diffusion layer
ϵ	dielectric constant
λ	logarithmic distribution coefficient
ρ	density
σ	standard deviation

Subject Index

Acetylacetone 36
Activated carbon 92
Airborne particulates 9, 108
Ammonium pyrrolidinedithiocarbamate 36
Ammonium tetramethylenedithiocarbamate 36
Andersen sampler 109
Anodic dissolution 74
Anodic stripping curve 69
APDC 36

Back-extraction 35
Backward-flow elution 79
Backwashing 34
Batch extraction 30
Batch operation 75
Berthelot-Nernst equation 58
Binomial distribution 33
Blank run 7
Blank value 7
Breakthrough curve 83

Carrier precipitate 61
Carrier precipitation 61, 95, 106
Cellulosic exchangers 88
Centrifugation 105, 108
Chelate extraction, equilibria 37
   of matrix elements 44
   systems 35
   of trace elements 43
Chelating resin 80, 81
Chromatographic column 77
Chromatographic elution curve 77
Class 100 cleanliness level 11
Clean benches 12
Clean hoods, balanced-laminar-airflow 12
   horizontal-laminar-airflow 12
   vertical-laminar-airflow 12
Clean rooms 10
   horizontal-laminar-airflow 11
   nonlaminar-airflow 11
   vertical-laminar-airflow 11
Closed-circuit distillation apparatus 23
Coextraction 42

Collector precipitates 61, 63, 95
Combustion method 27
Condensers 27
Conditioning 81
Consecutive stability constant 40
Container materials 13
   cleaning agents 16
   inorganic 13
   plastic 13
   selection of 13
   surface treatment 16
Contamination 6, 9
   airborne 9
   by analysts 20
   by apparatus 13
   during comminution 16
   by reagents 17
Continuous extraction 33
Continuous extractors 32
Controlled potential electrolysis 67
Coprecipitation 20, 57
Cupferron 36

DDTC 36
Decomposition techniques 3
Depolarizer 68
Desorption 75
Detection limits, absolute 2
   concentration 2
   relative 2
Determination techniques 3
Dialysis 104
Discontinuous countercurrent extraction 33, 43
Distillation 17, 21
   carrier 28
   isopiestic 19
   isothermal 19, 21
   sub-boiling 19
Distribution coefficient, effective 100
   equilibrium 100
   homogeneous 58
   logarithmic 58
   volume 75
   weight 75

Distribution ratio 30
Dithizone 36, 42
Doerner-Hoskins equation 58
Droplet countercurrent chromatography 34
Dry assay 55
Dry oxidation 28

Electrochemical deposition 20, 67, 107
Electrodeposition 67, 107
  of matrix elements 73
  on mercury cathodes 69
  on solid electrodes 67
  of trace elements 70
Electrolysis cells 67
  mercury cathode 71
  polyethylene 55
Electrostatic precipitation 108
Enrichment 4
  factor 6
Evaporation 21
  chambers 12
Extraction, of ion pairs 46
  of iron (III) chloride complexes 47
  of metal chelates 35
  of metal oxinates 41
  with molten organic compounds 51
  from nonaqueous samples 51
  from soil 55
Extraction constant 38
Extraction rate 43

Fajans-Paneth-Hahn rules 59
Filtration 20, 75, 104, 108
  apparatus 76
Fire assay 55
Floating-zone technique 101
Flotation 94
  cells 94
  general procedures 94
Freeze concentration 100
Freeze-drying 28

Gas evolution 21
Gaseous samples 108
Gathering precipitate 61
Gaussian normal distribution 33, 78
Gel filtration 104
Glassy carbon 14
Glove boxes 12

Half-extraction volume 33
HEPA filters 11
High-efficiency particulate air filters 11
High-purity reagents, laboratory preparation
  of 17
  selection of 17
Homogeneous distribution 58

Homogeneous extraction 50
Hot extraction 27
8-Hydroxyquinoline 36

Impactor, cascade 108
  multistage 108
Inert-gas fusion 27
In-line detectors 79
Inorganic ion exchangers 93
Inorganic trace analysis 1
  scheme of 2
Ion-association extraction, equilibria 47
  of matrix elements 50
  systems 46
  of trace elements 50
Ion exchange 19, 75, 107
  data 129
  equilibria 82
  resins 81
Ion flotation 98
Ion-retardation resin 81

Liquid chromatography 75, 107
Liquid-liquid extraction 20, 30, 106
  general procedures 30
Logarithmic distribution 58
Loss 5, 9, 20
  due to apparatus 13
Low-temperature ashing 29
Lyophilization 28

Macroreticular resin 91, 107
Masking 41, 59, 62
  agents 41, 42, 127
Microdiffusion unit 22
Microreticular resin 81
Microscale operations 8
Mixed crystal formation, anomalous 57
  isomorphous 57

Normal freezing 100

Occlusion 59
Organic solvents 126
Organic sorbents 90
Oxidation-reduction buffers 71
Oxine 36, 37
Oxygen bomb 29
Oxygen flask 29

PAN 36
Paper chromatography 79
Partial dissolution 53
Partition constant 30
Partition liquid chromatography 34
Pellicular ion exchangers 81
Percent extraction 31

Permeable sorbent disk   75
Plastic gloves   20
Plate theory   76
Platinum   14
Poisson distribution   78
Polyethylene   14, 15
Polyfluorocarbon   14
Polypropylene   14
Polyurethane foam   90
Precipitation, from homogeneous solution
  56
  of matrix elements   56
  of trace elements   61
Preconcentration   3
  coefficient   6
Pre-equilibration   31
Pressure bomb   26
Pyrex   14
1-(2-Pyridylazo)-2-naphthol   36
Pyrohydrolysis   28

8-Quinolinol   36

Radioactivation   6
Radioactive tracer technique   5
Rapidity   7
Reagent paper   91
Recrystallization   20, 60
Redox buffers   71
Reference electrode   67
Relative standard deviation   7
Reprecipitation   60
Reversed-phase partition chromatogra-
  phy   91, 107

Salting-out effect   50
Sample size   8
Sedimentation   108
Selective dissolution, of matrix   52
  of trace elements   53
Selectivity   62, 80
  coefficient   82
Separation, of gaseous trace constitu-
  ents   109
  of particles   104, 108
Simplicity   7
Snake-cage resin   81

Sodium diethyldithiocarbamate   36
Solid-liquid extraction   52
Solvent sublation   98
Sorption   75, 107
Specific capacity   80
Spontaneous electrochemical deposition   73
Standard samples   6
Stepwise formation constant   40
Stripping   35
Stripping voltammetry, anodic   69
  cathodic   69
Strong-acid cation exchange resin   80, 81
Strong-base anion exchange resin   80
Sublimation   21
Superficially porous ion exchanger   81
Surface adsorption   59
Surfactants   68, 96, 98
Synergism   42

Teflon   14, 15
Thenoyltrifluoroacetone   36
Theoretical plate   76
Thermal precipitation   108
Thin-layer chromatography   79
Three-phase extraction   50
Trace element analysis   1
Trace element speciation   103
Trace recovery   5
TTA   36

Ultrafiltration   104

Vacuum extraction   27
Vacuum fusion   27
Vitreous silica   14
Volatilization   21, 105
  of matrix   22, 28
  from solid and molten states   27
  from solution   22
  of trace elements   22, 27
Vycor   14

Water samples   44, 64, 84, 97, 99, 103
Wet oxidation   26

Zone melting   101

K. Cammann

Working with Ion-Selective Electrodes

Chemical Laboratory Practice

Translated from the German by A. H. Schroeder
1979. 65 figures, 8 tables. X, 226 pages. ISBN 3-540-09320-6

The field of ion-selective electrodes has grown enormously since the publication of the first edition of this work. The Second Edition, in English, considers new developments which have since taken place in gas sensors, enzyme electrodes and industrial applications of ion-selective electrodes.

Lucidly written and containing a helpful index, the book uses a new theoretical approach to explain the behavior of electrodes in a way comprehensible to a newcomer in the field. Various electrode types available are described as well as "do-it-yourself" electrodes, microelectrodes, industrial flow-thru assemblies, and pollution control monitors.

The accent of the book is on simple procedures. The user therefore learns to avoid errors resulting from an unfamiliarity with electrochemical measurements of single ion activities. The bibliography of the New Edition has been expanded considerably to include important new publications.

Already regarded as a standard among the literature in West Germany, this English Edition now offers analytical scientists worldwide a systematic and inviting introduction to the field of ion-selective electrodes.

H. Engelhardt

High Performance Liquid Chromatography

Chemical Laboratory Practice

Translated from the German by G. Gutnikov
1979. 73 figures, 13 tables. XII, 248 pages. ISBN 3-540-09005-3

This simple and non-mathematical introduction to high-performance-liquid chromatography (HPLC) emphasizes the practical aspects of achieving a successful separation. This method usually permits analyses to be carried out more rapidly than by gas chromatography and is moreover, eminently suited for the separation of heatlabile, high-boiling, or non-volatile substances, without lengthy or tedious derivatization. In principle, all substances that are stable in solution are amenable to separation by HPLC.

HPLC equipment is described in terms of the individual components, their expected performance capabilities and suitability for certain applications.

The areas of applications of the various separation techniques (adsorption, partition, ion-exchange, exclusion) are pointed out in order to facilitate selection of the most appropriate technique by the worker for his particular problem. Considerable discussion is devoted to the parameters that are important in optimizing or improving a given separation.

The application of HPLC to actual problems in organic chemistry, pharmacological research, medicine, biochemistry and petrochemistry are illustrated by numerous relevant examples. This book is a translation of the well-known and very successful German edition.

Springer-Verlag
Berlin
Heidelberg
New York

A. Maehly, L. Strömberg

Chemical Criminalistics

1981. 70 figures, 65 tables. VII, 322 pages
ISBN 3-540-10732-1

This book describes the chemical analyses carried out in forensic laboratories. More than just an essential reference for experts, it is also a study text, providing a systematic outline of forensic chemistry as well as examples of actual cases handled in the authors' laboratory. Some aspects of relatively sophisticated investigations, particularly in the drug field, are discussed. Organizational aspects of forensic science laboratories – also useful for the organization of other types of laboratories – are covered. Advice is given on how to locate pertinent literature to enable readers to deepen their knowledge of the field.

The book is concerned predominantly with the chemical aspects of forensic work. It will prove a valuable resource to forensic chemists and physicians, to prosecutors and defense lawyers, and to toxicologists and analytical chemists.

Springer-Verlag
Berlin
Heidelberg
New York

Contents: General Introduction: Historical Notes. Forensic Science Today. – The State of the Art: Narcotics and Dangerous Drugs. Explosives. Polymers. Fibers. Paints, Varnishes and Lacquers. Glass. Soil. Firearm Discharge Residues. Fire Investigation. Questioned Documents. Toxic Substances in Food. Restoration of Erased Markings. Miscellaneous. – Auxiliary Activities: The Forensic Significance of Physical Evidence and its Collection. Reference Collections. The Forensic Expert. Sources of Information on Forensic Science. The Organization of a Forensic Science Laboratory. – Index.